COMPACT DATA

FOR NAVIGATION AND ASTRONOMY

for the years

1986–1990

B. D. Yallop
and
C. Y. Hohenkerk

H.M. Nautical Almanac Office
Royal Greenwich Observatory

LONDON: HER MAJESTY'S STATIONERY OFFICE

First published 1985

ISBN 0 11 886926 4

PREFACE

This volume of *Compact Data for 1986–1990* is a continuation of the previous volume which was published in 1981 as *Royal Greenwich Observatory Bulletin* No 185 and covered the years 1981 to 1985.

The primary aim of this publication is to provide navigators and astronomers with simple and efficient methods for calculating the positions of the Sun, Moon, planets and navigational stars to low precision (about 0′·1) using a small programmable calculator or a personal computer.

A few changes and improvements have been introduced to this volume. For example, the Greenwich Hour Angle (*GHA*) and Declination (*DEC*) of the bright stars are each represented for the whole five-year period by one set of eight coefficients. The *GHA* and *DEC* may be evaluated on a monthly basis using these coefficients with a hand-held programmable calculator. More powerful computers may be programmed to calculate the *GHA* and *DEC* for any date during the whole five-year interval. A new table containing monthly Chebyshev coefficients for the Moon has been added for use with such computers. The *GHA*, *DEC* and horizontal parallax of the Moon may be calculated to better than 0′·2 using about 40 coefficients and a straightforward method of reduction.

Other minor changes to the main tables involve the representation of the semi-diameter of the Sun and the horizontal parallax of Venus and Mars by monthly coefficients which are easier to program than critical tables.

The explanation includes formulae for determining a position, without the use of sight-reduction tables, from observations made with a sextant. A simple guide on programming algebraic formulae has been added to help those who have recently acquired a calculator but are unfamiliar with such procedures.

This volume has been prepared at the Royal Greenwich Observatory, under the direction of the Superintendent of H M Nautical Almanac Office, Dr G A Wilkins. It was designed by Dr B D Yallop, who also wrote the Explanation; he was assisted in the design and preparation of the tables by Miss C Y Hohenkerk and in the production and proofreading by Mrs A F Strong and other members of the staff of the Office. Reproducible material for the main tabulations was prepared at the Rutherford and Appleton Laboratory with the assistance of Mrs K M Crennell.

ALEXANDER BOKSENBERG
Director

Royal Greenwich Observatory,
Herstmonceux Castle, Hailsham,
East Sussex, BN27 1RP, England

December 1984

RELATED PUBLICATIONS

The Astronomical Almanac contains precise heliocentric and geocentric coordinates of the Sun, Moon and planets. These annual volumes also contain ephemerides for natural satellites and some minor planets and for physical observations of the Sun, Moon and planets, as well as times of rising and setting of the Sun and Moon and data on stars and galaxies, the calendar and observatories.

Astronomical Phenomena is published annually in advance of *The Astronomical Almanac* from which it contains extracts giving the dates and times of planetary and lunar phenomena and other astronomical data of general interest.

The Nautical Almanac contains ephemerides at an interval of one hour and auxiliary astronomical data for marine navigation.

The Air Almanac is issued in two six-monthly parts and contains ephemerides at an interval of ten minutes and auxiliary astronomical data for air navigation. From 1987 onwards it will be issued annually.

Planetary and Lunar Coordinates 1984–2000 provides low-precision astronomical data for use in advance of the annual ephemerides and for other purposes. It contains heliocentric, geocentric, spherical and rectangular coordinates of the Sun, Moon and planets, eclipse data, and auxiliary data, such as orbital elements and precessional constants.

All the above publications are prepared jointly by H.M. Nautical Almanac Office, Royal Greenwich Observatory, and the Nautical Almanac Office, United States Naval Observatory, and are published jointly by Her Majesty's Stationery Office and the United States Government Printing Office.

The Star Almanac for Land Surveyors contains an ephemeris of the Sun, monthly positions of all stars brighter than magnitude 4·0, and other data of relevance to the determination of position and azimuth by astronomical techniques. It is prepared by H.M. Nautical Almanac Office and is published annually by Her Majesty's Stationery Office.

Compact Data for Navigation and Astronomy for 1981 to 1985 contains data, which are mainly in the form of polynomial coefficients, for use by navigators and astronomers to calculate the positions of the Sun, Moon, planets and bright stars using a simple pocket calculator. Formulae and methods are also given for calculating other related data with the aid of a programmable calculator. This booklet is available from the Royal Greenwich Observatory as R.G.O. Bulletin No. 185, price £3·30, including postage and packing.

CONTENTS

SUMMARY OF NOTATION AND RELATIONSHIPS

The following notation and relationships apply throughout this booklet.

GMT = Greenwich mean time (see note below)

D = the number of days elapsed from 1986 Jan 0 at 0^h GMT to the beginning of the month on day 0 at 0^h GMT. (See sections 4 and 5 and Table 5)

d = day of month

ϕ = latitude (north is positive)

λ = longitude (east is positive)

GHA = Greenwich hour angle

DEC = declination

SHA = sidereal hour angle

LHA = local hour angle

$GHA = GHA\ Aries + SHA$

$LHA = GHA + \lambda$

HP = horizontal parallax

S = semi-diameter

PA = parallax in altitude

H_S = sextant altitude

H = apparent altitude = sextant altitude corrected for instrumental error and dip

H_o = observed altitude = apparent altitude corrected for refraction and, if appropriate, corrected for parallax and semi-diameter

H_c = calculated altitude

Z = azimuth (true), measured clockwise around the horizon from north (0°) through east (90°), south (180°) and west (270°)

R = refraction

In this context the name Greenwich mean time (GMT) is synonymous with universal time (UT). Radio time signals are usually based on the system of coordinated universal time (UTC), which requires a small correction DUT of less than $0^s\!\cdot\!9$ to obtain UT, as follows,

$$UT = UTC + DUT$$

Sidereal hour angle (SHA) is related to right ascension (RA) by

$$SHA = 360° - RA$$

GHA Aries is synonymous with Greenwich sidereal time (GST).

Symbols are generally used to denote the physical quantitites which they represent rather than their numerical expressions in some particular units. Where a symbol is used for a numerical value, this is either specifically stated or the unit used is indicated after the symbol. Angular quantities are usually expressed in decimal notation in degrees or hours; for example, the longitude of Singapore (103° 50′ East) is given by $\lambda° = +103{\cdot}83$ or by $\lambda^h = +6{\cdot}922$, where $\lambda^h = \lambda°/15$.

EXPLANATION

1. INTRODUCTION

This booklet is intended primarily for use in the practice of astronomical navigation. Astronomers and others will find the data useful for calculations which require the positions of the Sun, Moon and the navigational planets and stars to low precision, as for example in calculations of times of rising and setting (section 8). The terminology is that used by navigators; astronomers may have to refer to the notation and relationships given on page vi. The positional accuracy for the Sun, planets and stars is around 2″, except for Venus near inferior conjunction and Mars near opposition, while for the Moon it is usually better than 0′·2. Table A contains a summary of the maximum errors when compared with the precise ephemeris. The maximum errors in *GHA* and *DEC* for those months when Venus is at inferior conjunction and Mars is at opposition are shown in Tables B and C.

Table A: Summary of Maximum Errors

Body	Argument	Period	No of terms	Maximum Error ′
Sun	*GHA*	monthly	5	0·03
	DEC		5	0·01
	S		2	0·03
Planets*	*GHA*	monthly	5	0·03
	DEC		5	0·02
Venus, Mars	*HP*		1 or 2	0·03
Moon	*GHA*	daily	3	0·29
	DEC		3	0·18
	HP		2	0·02
Moon	*GHA*	monthly	16 (max)	0·20
	DEC		16 (max)	0·20
	HP		9 (max)	0·20
Stars	*GHA*	5 yearly	4	0·04
	DEC		4	0·04
Aries (Mean)	*GHA Aries*	5 yearly	1	0·01

* Except Venus at inferior conjunction and Mars at opposition, see Tables B and C

Table B: Maximum Errors for Venus

Inferior Conjunction Date	Maximum Errors Month	GHA (′)	DEC (′)
1986 Nov 5	1986 Sept	0·05	0·03
	Oct	0·72	0·38
	Nov	0·29	0·51
	Dec	0·02	0·05
1988 June 13	1988 May	0·51	0·02
	June*	1·24	0·05
	July	0·19	0·10
1990 Jan 18	1989 Nov	0·04	0·01
	Dec	0·13	0·01
	1990 Jan*	1·62	0·05
	Feb	0·40	0·04

* In these months the *HP* of Venus may reach a maximum error of 0′·06.

Table C: Maximum Errors for Mars

Opposition Date	Maximum Errors Month	GHA (′)	DEC (′)
1986 July 10	1986 June	0·04	0·01
	July	0·12	0·01
	Aug	0·12	0·01
	Sept	0·05	0·00
1988 Sept 28	1988 Sept	0·11	0·05
	Oct	0·03	0·03
	Nov	0·05	0·02
1990 Nov 27	1990 Oct	0·05	0·01
	Nov	0·15	0·02

2. SUN AND PLANETS

2.1 *Monthly polynomial coefficients for the Sun and planets* Monthly polynomial coefficients a_0, a_1, a_2, a_3, a_4 are given in Table 1 for evaluating the Greenwich hour angle (*GHA*) and declination (*DEC*) of the Sun and the four navigational planets. Check sums are provided for checking that the coefficients have been entered into the calculator correctly. Each opening of the table contains data for four months.

The time variable x is calculated from

$$x = (d + GMT^{h}/24)/32$$

where d is the day of the month and GMT^{h} is the Greenwich mean time in hours.

$(GHA - GMT)$ in hours and DEC in degrees are calculated from expressions of the form

$$a_0 + a_1x + a_2x^2 + a_3x^3 + a_4x^4$$

GHA in degrees is obtained by adding GMT^h to $(GHA - GMT)^h$ and multiplying the result by 15 to convert from hours to degrees.

An efficient way of evaluating the polynomial expression, which only requires the basic operations of addition and multiplication, is to rewrite it in the form

$$(((a_4x + a_3)\,x + a_2)\,x + a_1)\,x + a_0$$

Start the calculation at the innermost set of parentheses as follows:

set $b_4 = a_4$ and evaluate successively

$$b_3 = b_4x + a_3 \quad b_2 = b_3x + a_2 \quad b_1 = b_2x + a_1 \quad b_0 = b_1x + a_0.$$

Then b_0 is the required value. Note that at each stage of the calculation only the previous value of b is required (see section 9 for further information on programming algebraic expressions).

Example Calculate the GHA, DEC and semi-diameter S of the Sun on 1986 March 17 at $13^h\,23^m\,25^s$ GMT.

$$d = 17 \quad GMT^h = 13{\cdot}3903 \quad x = (17 + 13{\cdot}3903 \div 24) \div 32 = 0{\cdot}548685$$

From Table 1 the coefficients for $(GHA - GMT)^h$ are

$a_4 = -0{\cdot}00501 \quad a_3 = -0{\cdot}01940 \quad a_2 = 0{\cdot}07032 \quad a_1 = 0{\cdot}09736 \quad a_0 = 11{\cdot}78841$

Hence $b_4 = -0{\cdot}00501$
$b_3 = -0{\cdot}00501 \times 0{\cdot}548685 - 0{\cdot}01940 = -0{\cdot}02215$
$b_2 = -0{\cdot}02215 \times 0{\cdot}548685 + 0{\cdot}07032 = 0{\cdot}05817$
$b_1 = 0{\cdot}05817 \times 0{\cdot}548685 + 0{\cdot}09736 = 0{\cdot}129275$
$b_0 = 0{\cdot}129275 \times 0{\cdot}548685 + 11{\cdot}78841 = 11{\cdot}859342$

$GHA^h = b_0 + GMT = 11{\cdot}8593 + 13{\cdot}3903 = 1{\cdot}2496$ (removing multiples of 24^h).

$GHA = 15 \times 1{\cdot}2496 = 18^\circ{\cdot}7443 = 18^\circ\,44'{\cdot}7$

From Table 1 the coefficients for DEC° are

$a_4 = 0{\cdot}0523 \quad a_3 = -0{\cdot}6787 \quad a_2 = 1{\cdot}0684 \quad a_1 = 12{\cdot}0589 \quad a_0 = -8{\cdot}1594$

Hence $b_4 = 0{\cdot}0523$
$b_3 = 0{\cdot}0523 \times 0{\cdot}548685 - 0{\cdot}6787 = -0{\cdot}6500$
$b_2 = -0{\cdot}6500 \times 0{\cdot}548685 + 1{\cdot}0684 = 0{\cdot}7118$
$b_1 = 0{\cdot}7118 \times 0{\cdot}548685 + 12{\cdot}0589 = 12{\cdot}4494$
$b_0 = 12{\cdot}4494 \times 0{\cdot}548685 - 8{\cdot}1594 = -1{\cdot}3286$

$DEC = b_0 = -1^\circ{\cdot}3286 = \text{S } 1^\circ\,19'{\cdot}7$

The semi-diameter S is calculated from the formula

$$S = a_0 + a_1x$$

From Table 1 $a_1 = -0{\cdot}0024 \quad a_0 = 0{\cdot}2695$

Hence $S = -0{\cdot}0024 \times 0{\cdot}548685 + 0{\cdot}2695 = 0^\circ{\cdot}2682 = 16'{\cdot}1$

2.2 *Corrections to the observed altitude of the Sun* The apparent altitude H of a limb of the Sun has to be corrected to the centre of the Sun. In addition a correction for parallax in altitude and for refraction must be applied. The observed altitude H_o is given by

$$H_o = H \pm S + HP \cos H - R$$

where S is the semi-diameter, $HP = 0^\circ{\cdot}0024$ is the horizontal parallax and R is the refraction. The plus sign is used if the lower limb was observed. The minus sign is used if the upper limb was observed.

2.3 *Corrections to the observed altitudes of the planets* Corrections have been incorporated in the data for Venus and Mars to allow for their phase, so that the position obtained from the polynomial coefficients in Table 1 is for the observed centre-of-light.

For Venus and Mars a correction for parallax in altitude has to be applied, and for all the planets a correction for refraction is required. The observed altitude for Venus and Mars is given by

$$H_o = H + HP \cos H - R$$

where HP is the horizontal parallax and R is the refraction. Monthly coefficients for HP are given at the foot of each page of Table 1 where

$$HP = a_0 + a_1 x$$

and $$x = (d + GMT^{\mathrm{h}}/24)/32$$

Section 6 gives formulae for calculating refraction.

3. MOON

The GHA, DEC and HP of the Moon may either be calculated from the daily polynomial coefficients in Table 2 or from the monthly Chebyshev coefficients in Table 3. Since the Chebyshev coefficients are more difficult to use it may be advisable to ignore section 3.2 on the first reading.

3.1 *Daily polynomial coefficients for the Moon* Table 2 gives daily polynomial coefficients a_0, a_1, a_2 to calculate the Moon's Greenwich hour angle (GHA) and declination (DEC), both in degrees, and daily polynomial coefficients a_0, a_1 to calculate horizontal parallax HP, also in degrees. Each consecutive pair of pages in Table 2 contains data for three months.

The time variable x is calculated from

$$x = GMT^{\mathrm{h}}/24$$

where GMT^{h} is the Greenwich mean time in hours. The GHA and DEC in degrees are calculated from an expression of the form

$$a_0 + a_1 x + a_2 x^2$$

and the *HP* from an expression of the form

$$a_0 + a_1 x$$

Example Calculate *GHA*, *DEC* and *HP* of the Moon on 1986 October 20 at $5^h\ 47^m\ 22^s$ GMT.

$$GMT^h = 5{\cdot}7894 \quad x = 5{\cdot}7894 \div 24 = 0{\cdot}241227$$

The coefficients for *GHA*° from Table 2 are

$$a_2 = -0{\cdot}2428 \quad a_1 = 348{\cdot}7309 \quad a_0 = 340{\cdot}0467$$

Hence $GHA = (-0{\cdot}2428 \times 0{\cdot}241227 + 348{\cdot}7309) \times 0{\cdot}241227 + 340{\cdot}0467$

$= 64°{\cdot}156 = 64°\ 09'{\cdot}3$ (removing multiples of 360°)

The coefficients for *DEC*° from Table 2 are

$$a_2 = -0{\cdot}5106 \quad a_1 = 4{\cdot}1834 \quad a_0 = 20{\cdot}6673$$

Hence $DEC = (-0{\cdot}5106 \times 0{\cdot}241227 + 4{\cdot}1834) \times 0{\cdot}241227 + 20{\cdot}6673$

$= +21°{\cdot}647 = \text{N}\ 21°\ 38'{\cdot}8$

The coefficients for *HP*° from Table 2 are

$$a_1 = -0{\cdot}0057 \quad a_0 = 0{\cdot}9137$$

Hence $HP = -0{\cdot}0057 \times 0{\cdot}241227 + 0{\cdot}9137 = 0°{\cdot}912 = 54'{\cdot}7$

3.2 *Monthly Chebyshev coefficients for the Moon* Table 3 contains *n* Chebyshev polynomial coefficients $a_0, a_1, a_2, \ldots a_{n-1}$ for calculating λ, β and *HP* for periods of one month, where λ and β are the ecliptic longitude and latitude and *HP* is the horizontal parallax. Check sums are provided for checking that the coefficients have been entered into the calculator or computer correctly. For λ and β the maximum number of coefficients is 16, and for *HP* the maximum is 9.

The Moon's right ascension (*RA*) and declination (*DEC*) are derived from trigonometric expressions involving λ, β and the adopted value of the obliquity $\varepsilon = 23°{\cdot}44$.

Each page of Table 3 contains data for six months. The time variable *x* is calculated from

$$x = (d + GMT^h/24)/32$$

where *d* is the day of the month and GMT^h is the Greenwich mean time in hours.

Each of the quantities λ, β and *HP* is derived as follows:

Calculate $y = 2(2x - 1)$

Set $b_{n+1} = b_n = 0$

Use the recurrence relation

$$b_i = yb_{i+1} - b_{i+2} + a_i$$

for $i = n - 1, n - 2, \ldots, 2, 1, 0$ to calculate b_2 and b_0. Then the required quantity, in degrees, is obtained from the expression

$$(b_0 - b_2 + a_0)/2$$

The Moon's RA and DEC are calculated as follows:

Set $X = \cos\beta \cos\lambda \quad Y = \cos\varepsilon \cos\beta \sin\lambda - \sin\varepsilon \sin\beta$

and $Z = \sin\varepsilon \cos\beta \sin\lambda + \cos\varepsilon \sin\beta$

where $\varepsilon = 23^\circ{\cdot}44$

Then $RA^\circ = \tan^{-1}(Y/X)$ and $DEC^\circ = \sin^{-1}Z$

If $X < 0$ add 180° to RA. If $X > 0$ and $Y < 0$ add 360° to RA.

To obtain the Greenwich hour angle of the Moon (GHA) calculate

$$L^\circ = 0{\cdot}9856474(D + d + GMT^{h}/24)$$

where GMT^{h} is the Greenwich mean time in hours, d is the day of the month and D is the number of days elapsed from 1986 Jan 0 at 0^{h} GMT to the beginning of the month on day 0 at 0^{h} GMT. D may be obtained from Table 5. Alternatively it may be calculated using the formula in section 4.2.

Then $GHA\ Aries = 99{\cdot}3672 + L^\circ + 15\ GMT^{h}$

and $GHA = GHA\ Aries - RA$

The following remarks on the recurrence relation should be noted:

1. The total number of Chebyshev coefficients is n but the highest order coefficient is a_{n-1}.

2. The time variable x is in the range 0 to 1, so that y is in the range -2 to $+2$.

3. The expression $(b_0 - b_2 + a_0)/2$ is also equal to $yb_1/2 - b_2 + a_0$ since $b_0 = yb_1 - b_2 + a_0$. If this latter expression is used and the recurrence relation is stopped at $i = 1$ instead of $i = 0$ then at any stage of the calculation only two successive values of b are required. This modification allows a slight but valuable saving of storage space in the calculator.

Example Using the monthly Chebyshev coefficients calculate the Moon's GHA, DEC and HP on 1986 October 20 at $5^{h}\ 47^{m}\ 22^{s}$ GMT.

$$GMT^{h} = 5{\cdot}7894 \quad d = 20 \quad x = 0{\cdot}632538 \quad y = 0{\cdot}530153$$

For longitude λ, from Table 3, $n = 12$

Hence

$$\begin{aligned}
b_{13} &= 0 \\
b_{12} &= 0 \\
b_{11} &= yb_{12} - b_{13} + a_{11} \\
&= 0{\cdot}530153 \times 0{\cdot}0 - 0{\cdot}0 - 0{\cdot}0049 = -0{\cdot}0049 \\
b_{10} &= yb_{11} - b_{12} + a_{10} \\
&= 0{\cdot}530153 \times (-0{\cdot}0049) - 0{\cdot}0 + 0{\cdot}0067 = +0{\cdot}0041 \\
b_9 &= yb_{10} - b_{11} + a_9 \\
&= 0{\cdot}530153 \times 0{\cdot}0041 - (-0{\cdot}0049) + 0{\cdot}0109 = +0{\cdot}0180 \\
b_8 &= yb_9 - b_{10} + a_8 \\
&= 0{\cdot}530153 \times 0{\cdot}0180 - 0{\cdot}0041 - 0{\cdot}0523 = -0{\cdot}0469 \\
b_7 &= yb_8 - b_9 + a_7 \\
&= 0{\cdot}530153 \times (-0{\cdot}0469) - 0{\cdot}0180 - 0{\cdot}0175 = -0{\cdot}0603
\end{aligned}$$

$$b_6 = yb_7 - b_8 + a_6$$
$$= 0{\cdot}530153 \times (-0{\cdot}0603) - (-0{\cdot}0469) - 0{\cdot}0912 = -0{\cdot}0763$$
$$b_5 = yb_6 - b_7 + a_5$$
$$= 0{\cdot}530153 \times (-0{\cdot}0763) - (-0{\cdot}0603) - 0{\cdot}0152 = +0{\cdot}0047$$
$$b_4 = yb_5 - b_6 + a_4$$
$$= 0{\cdot}530153 \times 0{\cdot}0047 - (-0{\cdot}0763) + 2{\cdot}4185 = +2{\cdot}4973$$
$$b_3 = yb_4 - b_5 + a_3$$
$$= 0{\cdot}530153 \times 2{\cdot}4973 - 0{\cdot}0047 + 0{\cdot}6654 = +1{\cdot}9847$$
$$b_2 = yb_3 - b_4 + a_2$$
$$= 0{\cdot}530153 \times 1{\cdot}9847 - 2{\cdot}4973 - 5{\cdot}0678 = -6{\cdot}5129$$
$$b_1 = yb_2 - b_3 + a_1$$
$$= 0{\cdot}530153 \times (-6{\cdot}5129) - 1{\cdot}9847 + 210{\cdot}1379 = +204{\cdot}7004$$
$$b_0 = yb_1 - b_2 + a_0$$
$$= 0{\cdot}530153 \times 204{\cdot}7004 - (-6{\cdot}5129) + 353{\cdot}4507 = +468{\cdot}4862$$

Hence $\lambda = (b_0 - b_2 + a_0)/2 = (468{\cdot}4862 + 6{\cdot}5129 + 353{\cdot}4507) \div 2$
$= 54^\circ{\cdot}2249$ (removing multiples of 360°).

A similar calculation for latitude β with $n = 11, b_{12} = b_{11} = 0$

yields $b_2 = -0{\cdot}2612 \quad b_0 = +4{\cdot}3747$
$\beta = (b_0 - b_2 + a_0)/2 = 2^\circ{\cdot}9074$

Finally for HP with $n = 7, b_9 = b_8 = 0$

yields $b_2 = +0{\cdot}03455 \quad b_0 = +0{\cdot}90302$
$HP = (b_0 - b_2 + a_0)/2$
$= (0{\cdot}90302 - 0{\cdot}03455 + 0{\cdot}9536) \div 2 = 0^\circ{\cdot}911 = 54'{\cdot}7$

Since $\lambda = 54^\circ{\cdot}2249, \beta = 2^\circ{\cdot}9074$
$X = \cos\beta\cos\lambda = 0{\cdot}998713 \times 0{\cdot}584605 = 0{\cdot}583853$
$Y = \cos\varepsilon\cos\beta\sin\lambda - \sin\varepsilon\sin\beta$
$= 0{\cdot}917477 \times 0{\cdot}998713 \times 0{\cdot}811318 - 0{\cdot}397789 \times 0{\cdot}050722 = 0{\cdot}723231$
$Z = \sin\varepsilon\cos\beta\sin\lambda + \cos\varepsilon\sin\beta$
$= 0{\cdot}397789 \times 0{\cdot}998713 \times 0{\cdot}811318 + 0{\cdot}917477 \times 0{\cdot}050722 = 0{\cdot}368854$

Hence $RA = \tan^{-1}(Y/X) = \tan^{-1}(0{\cdot}723231/0{\cdot}583853) = 51^\circ{\cdot}0866$
$DEC = \sin^{-1}Z = \sin^{-1}(0{\cdot}368854) = +21^\circ{\cdot}6449 = \text{N } 21^\circ\ 38'{\cdot}7$

From Table 5, for 1986 October, $D = 273$

Hence $L = 0{\cdot}9856474(273 + 20 + 5{\cdot}7894/24) = 289{\cdot}0325$
$GHA\ Aries = 99{\cdot}3672 + 289{\cdot}0325 + 15 \times 5{\cdot}7894 = 115{\cdot}2413$
(removing multiples of 360°)
$GHA = GHA\ Aries - RA = 115^\circ{\cdot}2413 - 51^\circ{\cdot}0866$
$= 64^\circ{\cdot}1547 = 64^\circ\ 09'{\cdot}3$

3.3 *Corrections to the observed altitude of the Moon* The apparent altitude H of a limb of the Moon has to be corrected to the centre of the Moon. In addition corrections for parallax and refraction have to be applied. The observed altitude H_o is given by

$$H_o = H \pm S + HP\cos H - R$$

where HP is the horizontal parallax in degrees, $S = 0°{\cdot}2724\ HP$ is the semi-diameter of the Moon and R is the refraction (see section 6 for formulae for refraction). The plus sign for S is used if the lower limb was observed. The minus sign for S is used if the upper limb was observed.

The correction for parallax does not allow for the oblateness of the Earth which could produce an error of up to $0'{\cdot}2$. To reduce this error a further correction of

$$-0°{\cdot}0032 \sin^2 \phi \cos H + 0°{\cdot}0032 \sin 2\,\phi \cos Z \sin H$$

should be added where ϕ is the latitude and Z is the azimuth of the Moon. At mid-latitudes and for altitudes below 60° a simple approximation to this expression is $-0°{\cdot}0017 \cos H$.

4. AUXILIARY DATA

4.1 *Phases of Moon and Seasons* Table 4 gives the phases of the Moon to the nearest minute and the seasons.

4.2 *GHA Aries* Table 5 may be used to calculate *GHA Aries* which is equivalent to Greenwich mean sidereal time (*GMST*). Calculate

$$L° = 0{\cdot}9856474\,(D + d + GMT^{\mathrm{h}}/24)$$

where GMT^{h} is the Greenwich mean time in hours, d is the day of the month and D is the number of days elapsed from 1986 Jan 0 at 0^{h} to the beginning of the month at day 0 at 0^{h}. D may be obtained from Table 5 or it may be calculated from the expression

$$D = \mathrm{INT}(365{\cdot}25y) + (\mathrm{INT}((y - 4\mathrm{INT}(y/4) + 3)/4) - 1) + \mathrm{INT}(275m/9) - (1 + \mathrm{INT}((y - 4\mathrm{INT}(y/4) + 3)/4))\,\mathrm{INT}((m + 9)/12) - 725416$$

where y is the year, m is the month (January = 1, February = 2 etc) and INT(x) means take the integer part of x.

Then $\quad GHA\ Aries = 99{\cdot}3672 + L° + 15\ GMT^{\mathrm{h}}$

The Greenwich apparent sidereal time *GAST* is obtained by adding the equation of the equinoxes (*NUT*) to *GMST* where

$$NUT = -(0°{\cdot}0048 \sin(35°{\cdot}9 - 0{\cdot}05295\,t) + 0°{\cdot}0004 \sin(198°{\cdot}7 + 1{\cdot}97129t)) \cos \varepsilon$$

and $\quad t = D + d + GMT^{\mathrm{h}}/24 \quad$ and $\varepsilon = 23°{\cdot}44$

Hence $\quad GAST = GHA\ Aries + NUT$

5. STARS

5.1 *Bright Stars* Coefficients a_0, a_1, a_2, a_3 covering the whole 5-year period from 1986 to 1990 are given in Table 6 for evaluating the Greenwich hour angle (*GHA*) and declination (*DEC*) of the bright stars used in navigation. The stars are identified in Table 6 by their navigational number and their proper name. The table also contains the Bayer letter and constellation of each star and its magnitude (Mag).

The *GHA* and *DEC* for the stars are calculated from expressions of the form

$$GHA^\circ = a_0 + a_1 L + a_2 \sin L + a_3 \cos L + 15\, GMT^h$$
$$DEC^\circ = a_0 + a_1 L + a_2 \sin L + a_3 \cos L$$

where $L^\circ = 0{\cdot}9856474(D + d + GMT^h/24)$

D is the number of days elapsed from 1986 January 0 at 0^h to the beginning of the month on day 0 at 0^h, d is the day of the month and GMT^h is the Greenwich mean time in hours. Values of D for each month are given in Table 5. Alternatively D may be calculated from the expression in section 4.2.

Example Calculate the *GHA* and *DEC* of *Spica* (No 33) on 1989 July 17 at $19^h\ 27^m\ 33^s$ GMT

$$d = 17 \quad GMT^h = 19{\cdot}4592$$

From Table 5, for 1989 July, $D = 1277$. Alternatively using the expression for D in section 4,

$$y = 1989 \quad m = 7 \quad d = 17$$

$$\begin{aligned}
&\mathrm{INT}(365{\cdot}25y) = \mathrm{INT}(726482{\cdot}25) = 726482\\
&\mathrm{INT}(y/4) = \mathrm{INT}(1989/4) = \mathrm{INT}(497{\cdot}25) = 497\\
&\mathrm{INT}((y - 4\mathrm{INT}(y/4) + 3)/4) = \mathrm{INT}((1989 - 1988 + 3)/4) = 1\\
&\mathrm{INT}(275m/9) = \mathrm{INT}(213{\cdot}8889) = 213\\
&\mathrm{INT}((m + 9)/12) = \mathrm{INT}(1{\cdot}3333) = 1\\
&D = 726482 + (1 - 1) + 213 - (1 + 1) \times 1 - 725416 = 1277
\end{aligned}$$

Hence $L = 0{\cdot}9856474 \times (1277 + 17 + 19{\cdot}4592 \div 24) = 1276{\cdot}2269$

From Table 6, for *Spica* the coefficients for GHA° are

$$a_0 = 258{\cdot}2543 \quad a_1 = 0{\cdot}9999631 \quad a_2 = -0{\cdot}0052 \quad a_3 = 0{\cdot}0012$$

Hence
$$\begin{aligned}
GHA &= 258{\cdot}2543 + 0{\cdot}9999631 \times 1276{\cdot}2269 - 0{\cdot}0052 \times \sin 1276{\cdot}2269\\
&\quad + 0{\cdot}0012 \times \cos 1276{\cdot}2269 + 15 \times 19{\cdot}4592\\
&= +1826^\circ{\cdot}3219 \text{ (remove multiples of } 360^\circ),
\end{aligned}$$

Then $GHA = 26^\circ{\cdot}3219 = 26^\circ\ 19'{\cdot}3$

From Table 6, for *Spica* the coefficients for DEC° are

$$a_0 = -11{\cdot}0884 \quad a_1 = -0{\cdot}0000158 \quad a_2 = -0{\cdot}0020 \quad a_3 = 0{\cdot}0007$$

Hence
$$\begin{aligned}
DEC &= -11{\cdot}0884 - 0{\cdot}0000158 \times 1276{\cdot}2269 - 0{\cdot}0020 \times \sin 1276{\cdot}2269\\
&\quad + 0{\cdot}0007 \times \cos 1276{\cdot}2269\\
&= -11^\circ{\cdot}1087 = \mathrm{S}\ 11^\circ\ 06'{\cdot}5
\end{aligned}$$

5.2 *Circumpolar stars* Data for the circumpolar stars, *Polaris* (No 58) and *σ Octantis* (No 59), are also included at the end of Table 6. The coefficients give *GHA* and polar distance (p) which are calculated from the expressions

$$GHA = a_0 + a_1 L + a_2 \sin L + a_3 \cos L + 15\, GMT^h$$
$$p = a_0 + a_1 L + a_2 \sin L + a_3 \cos L$$

where L is calculated in the same way as for the non-circumpolar stars (see section 5).

Example Calculate GHA and p for *Polaris* on 1986 May 16 at $4^h\ 42^m$ GMT

$d = 16 \quad GMT^h = 4{\cdot}7$ From Table 5, for 1986 May, $D = 120$

Hence $L = 0{\cdot}9856474 \times (120 + 16 + 4{\cdot}7 \div 24) = 134{\cdot}2411$

From Table 6 for *Polaris* the coefficients for GHA are

$a_0 = 65{\cdot}2371 \quad a_1 = 0{\cdot}9992240 \quad a_2 = 0{\cdot}3466 \quad a_4 = -0{\cdot}1882$

Hence $GHA = 65{\cdot}2371 + 0{\cdot}9992240 \times 134{\cdot}2411 + 0{\cdot}3466 \times \sin 134{\cdot}2411$
$- 0{\cdot}1882 \times \cos 134{\cdot}2411 + 15 \times 4{\cdot}7$
$= 270^\circ{\cdot}2536 = 270^\circ\ 15'{\cdot}2$

From Table 6 for *Polaris* the coefficients for p are

$a_0 = 0{\cdot}7983 \quad a_1 = -0{\cdot}0000137 \quad a_2 = -0{\cdot}0022 \quad a_3 = -0{\cdot}0050$

Hence $p = 0{\cdot}7983 - 0{\cdot}0000137 \times 134{\cdot}2411 - 0{\cdot}0022 \times \sin 134{\cdot}2411$
$- 0{\cdot}0050 \times \cos 134{\cdot}2411$
$= 0^\circ{\cdot}7984 = 47'{\cdot}9$

5.3 *Use of circumpolar stars to determine latitude and azimuth* The following procedure may be used to determine latitude and azimuth from the observed altitude H_o of the circumpolar stars.

From Tables 5 and 6 calculate GHA and p as shown in section 5.2.

Use the longitude λ to calculate the local hour angle LHA from

$$LHA = GHA + \lambda$$

Form the quantities C and S from

$$C = p \cos(LHA) \quad S = p \sin(LHA)$$

Then latitude $= H_o - C + 0{\cdot}0087\, S^2 \tan H_o$

and azimuth of *Polaris* $= -S / \cos H_o$

azimuth of σ *Octantis* $= 180 + S / \cos H_o$

If the azimuth for *Polaris* is negative add 360°.

Example Calculate the observer's latitude and the azimuth of *Polaris* from an observation made in 1986 May 16 at $4^h\ 42^m$ GMT when the observer's longitude is assumed to be W 138°·376 and the observed altitude of *Polaris* is 35°·365.

From the previous example $GHA = 270^\circ{\cdot}2536$, $p = 0^\circ{\cdot}7984$

Hence $LHA = GHA - 138{\cdot}376 = 131{\cdot}8776$

$C = 0{\cdot}7984 \times \cos 131{\cdot}8776 = -0{\cdot}5329$

$S = 0{\cdot}7984 \times \sin 131{\cdot}8776 = 0{\cdot}5944$

latitude $= 35{\cdot}365 - (-0{\cdot}5329) + 0{\cdot}0087 \times 0{\cdot}5944 \times 0{\cdot}5944 \times 0{\cdot}7097$
$= +35^\circ{\cdot}900 = \text{N } 35^\circ\ 54'{\cdot}0$

azimuth $= -0{\cdot}5944 / 0{\cdot}8155 = 359^\circ{\cdot}271$

6. REFRACTION AND DIP

The refraction correction R_0, in degrees, for a standard temperature of 10°C and pressure of 1010 mb, for altitudes greater than 15° may be represented by

$$R_0 = 0^\circ{\cdot}0162 / \tan H$$

where H is the apparent altitude in degrees. For altitudes less than 15°, R_0 may be calculated from the expression

$$R_0 = (0{\cdot}5743 + 0{\cdot}0705\, H + 0{\cdot}00007\, H^2)/(1 + 0{\cdot}505\, H + 0{\cdot}0845\, H^2)$$

Alternatively R_0 may be calculated for any altitude from a single formula (see paper by G G Bennett, 1982, *J. Inst. Nav.*, vol **35**, p 255)

$$R_0 = 0^\circ{\cdot}0167 / \tan(H + 7{\cdot}31/(H + 4{\cdot}4))$$

An approximate expression for the correction f for a non-standard temperature T in degrees Celsius (°C) and pressure in millibars (mb) is given by

$$f = 0{\cdot}28\, P/(T + 273)$$

The total refraction $R = R_0 f$ should be subtracted from the apparent altitude H to give the observed altitude H_o.

If a marine sextant is used a correction D_h in degrees for the dip of the horizon must be subtracted from the sextant altitude. D_h is given by

$$D_h = 0^\circ{\cdot}0293\, h^{\frac{1}{2}}$$

where h is the height of the observer above the horizon in metres ($h^{\frac{1}{2}}$ is the square root of h).

For a bubble sextant $D_h = 0$.

7. SIGHT REDUCTION FORMULAE

The formulae in this section are intended for the determination of position at sea from observations made with a marine sextant. If the observations are made with a bubble sextant then no correction for dip is required.

First the sextant altitude is reduced to the observed altitude by adding the corrections for dip, index error, refraction, semi-diameter for the Sun and Moon and parallax for the Moon, Venus and Mars. The altitude and azimuth are calculated from the *GHA* and *DEC* of the body at the time of observation using the appropriate polynomial coefficients in this booklet. The position is obtained either by plotting the position lines using the intercept and azimuth method or by direct calculation using a method described below which is based on least squares.

7.1 *Reduction from sextant altitude to observed altitude* The observed altitude is obtained from the sextant altitude as follows:

1. Calculate dip

$$D_h = 0^\circ{\cdot}0293\, h^{\frac{1}{2}}$$

where h is the height of eye above the horizon in metres.

2. Calculate the apparent altitude

$$H = H_s + I - D_h$$

where H_s is the sextant altitude and I is the sextant index error.

3. Calculate refraction

$$R = 0^\circ{\cdot}0162 / \tan H$$

(see section 6 for alternative formulae for refraction)

4a. For the Sun calculate the semi-diameter S from Table 1.

4b. Also for the Sun calculate the parallax in altitude

$$PA = 0^\circ{\cdot}0024 \cos H$$

5a. For the Moon calculate the Moon's semi-diameter from

$$S = 0^\circ{\cdot}2724\ HP$$

where HP is the horizontal parallax which is obtained from Table 2.

5b. For the Moon calculate the parallax in altitude

$$PA = HP \cos H$$

5c. For the Moon apply a correction to the parallax for the oblateness of the Earth of

$$-0^\circ{\cdot}0032 \sin^2 \phi \cos H + 0^\circ{\cdot}0032 \sin 2\phi \cos Z \sin H$$

where ϕ is the latitude of the observer and Z is the azimuth of the Moon (see section 7.2).

6. For Venus and Mars calculate the parallax in altitude

$$PA = HP \cos H$$

where HP is the horizontal parallax which is obtained from Table 1, and set $S = 0$.

For the other planets and for stars, set $S = 0$ and $PA = 0$

7. Calculate the observed altitude

$$H_o = H \pm S + PA - R$$

where the plus sign is used if the lower limb of the Sun and Moon was observed and the minus sign if the upper limb was observed.

7.2 *The calculated altitude and azimuth* The altitude and azimuth (H_c, Z) are calculated from the GHA and DEC of the body and the estimated longitude and latitude (λ, ϕ) as follows:

1. Calculate the local hour angle

$$LHA = GHA + \lambda$$

2. The calculated altitude (H_c) is given by

$$H_c = \sin^{-1}(\sin \phi \sin DEC + \cos \phi \cos DEC \cos LHA)$$

3. The true azimuth (Z) is calculated from

$$X = \cos\phi \sin DEC - \sin\phi \cos DEC \cos LHA$$
$$Y = -\cos DEC \sin LHA$$
$$A = \tan^{-1}(Y/X)$$

If $X < 0$ then $Z = A + 180°$
If $X > 0$ and $Y < 0$ then $Z = A + 360°$
Else $Z = A$.

To avoid division by zero at the E and W cardinal points, which may produce an error in the calculator, add a very small number to the denominator X before division (eg $1{\cdot}0 \times 10^{-10} \equiv 1{\cdot}0\text{E}{-}10$).

Many calculators have the facility to convert directly from rectangular coordinates (x, y) to polar coordinates (r, θ). Set $x = X$, $y = Y$ and convert to polar coordinates. Then $Z = \theta$.

7.3 *Position from intercept and azimuth using position lines* The position line for an observation is determined from the intercept $p = H_o - H_c$ and the azimuth Z as follows:

1. Make an initial estimate of the position (λ_f, ϕ_f) at the time of fix (t_f).

2. Determine the position (λ, ϕ) at the time of observation (t) by dead reckoning (DR) from the time of fix. For example, if the bearing (or course) T and the speed V (in knots) remain constant

then
$$\lambda = \lambda_f + (t - t_f)(V/60) \sin T/\cos\phi_f$$
$$\phi = \phi_f + (t - t_f)(V/60) \cos T$$

where time is measured in hours.

3. Using the position at the time of observation (λ, ϕ) and the formulae in section 7.2 calculate the altitude H_c and azimuth Z. Calculate the intercept

$$p = H_o - H_c$$

4. The equation of a position line with origin at the estimated position at the time of fix (λ_f, ϕ_f) is given by

$$x \sin Z + y \cos Z = p$$

where the x-axis is parallel to lines of constant latitude, positive to the east, and the y-axis is along the meridian, positive to the north. If p is expressed in nautical miles (nm) then x and y will be in nautical miles also. Then the position line may be drawn on a chart in the standard way using the intercept and azimuth p, Z.

This procedure is repeated for each observation. If the errors are small the position will lie near the points of intersection of the position lines so that at least two position lines are required to obtain a fix.

7.4 *Position from intercept and azimuth by calculation* Alternatively the position of the fix may be calculated directly from two or more sextant observations, by using the method of least squares as follows:

If p_1, Z_1 are the intercept and azimuth of the first observation, p_2, Z_2 of the second observation and so on, form the quantities

$$\begin{aligned}
A &= \cos^2 Z_1 + \cos^2 Z_2 + \cdots \\
B &= \cos Z_1 \sin Z_1 + \cos Z_2 \sin Z_2 + \cdots \\
C &= \sin^2 Z_1 + \sin^2 Z_2 + \cdots \\
D &= p_1 \cos Z_1 + p_2 \cos Z_2 + \cdots \\
E &= p_1 \sin Z_1 + p_2 \sin Z_2 + \cdots \\
F &= p_1^2 + p_2^2 + \cdots
\end{aligned}$$

As a check verify that $A + C = n$ where n is the total number of observations included in the solution.

Calculate $G = AC - B^2$

Then an improved estimate of the position at the time of fix is

$$\lambda_f + d\lambda, \quad \phi_f + d\phi$$

where $d\lambda = (AE - BD)/(G \cos \phi_f)$ and $d\phi = (DC - EB)/G$

Additional observations may be included in the solution by simply adding the extra terms to A, B, C, D and E and calculating $d\lambda$ and $d\phi$ again. Similarly, observations may be rejected by subtracting the terms that were originally added to A, B, C, D and E and calculating $d\lambda$ and $d\phi$ again.

If the initial position λ_f, ϕ_f is not well known it will be necessary to iterate until a consistent solution is found. At each iteration λ_f and ϕ_f are updated to the latest calculated position. Indeed it is possible to start from any position on the Earth, and, provided that λ_f is kept in the range $-180°$ to $+180°$ and ϕ_f in the range $-90°$ to $+90°$, the solution in most cases will begin to converge after a few iterations.

7.5 *Estimated position error* If three or more position lines are obtained an estimate of the error in position may be calculated. The standard deviation of the estimated position σ in nautical miles is given by

$$\sigma = 60(S/(n-2))^{\frac{1}{2}}$$

where $S = F - D\,d\phi - E \cos \phi_f\,d\lambda$

The standard deviations σ_λ, σ_ϕ in longitude and latitude are given by

$$\sigma_\lambda = \sigma(A/G)^{\frac{1}{2}} \quad \sigma_\phi = \sigma(C/G)^{\frac{1}{2}}$$

In general as the number of observations increases the error in the estimated position decreases. Statistical theory shows that the estimated position has a probability P of lying within a confidence ellipse which is specified by the lengths of its axes a and b and the azimuth θ of the a-axis, where

$$\begin{aligned}
\tan 2\theta &= 2B/(A - C) \\
a &= \sigma k/(n/2 + B/\sin 2\theta)^{\frac{1}{2}} \\
b &= \sigma k/(n/2 - B/\sin 2\theta)^{\frac{1}{2}}
\end{aligned}$$

and $k = (-2 \log_e(1 - P))^{\frac{1}{2}}$ is a scale factor.

Values of the scale factor k for selected values of P are given in the table below.

Probability, P	0·39	0·50	0·75	0·90	0·95
Scale factor, k	1·0	1·2	1·7	2·1	2·4

Normally a confidence level of 95% is chosen, that is $P = 0{\cdot}95$.

The shape of the confidence ellipse depends only upon n and the distribution of the observations in azimuth; whilst the size of the ellipse, apart from the scale factor, depends upon the errors of observation. The method assumes that the observations have equal weight. The best results will be obtained when the observations are equally spaced in azimuth. In such cases the effect of systematic errors on the final calculated position will be minimised.

7.6 *Plotting position lines and the confidence ellipse with a personal computer*
Personal computers usually have facilities for plotting either on a TV monitor screen or on some other plotting device. They normally use the principle that a cursor or pen is moved from a point A with rectangular coordinates x_A, y_A to a point B with rectangular coordinates x_B, y_B and a straight line is drawn joining A and B in the process. A curve, such as an ellipse, is drawn by approximating it to a succession of short straight lines; the curve looks smooth to the eye provided that the lines are short enough.

The origin for plotting is chosen initially to be the estimated position of the fix with the x-axis along the line of latitude, positive to the east, and the y-axis along the meridian, positive to the north.

The position lines and confidence ellipse are drawn inside a square of length 20 nautical miles centred on the estimated position at the time of fix with sides parallel to the x and y axes.

To draw the position line for intercept p (in nautical miles) and azimuth Z first find the two points where this line crosses the sides of the square. Set $x = \pm 10$. Then $y = (p \mp 10 \sin Z)/\cos Z$. The line crosses the side of the square if $-10 \leqslant y \leqslant 10$. Similarly set $y = \pm 10$. Then $x = (p \mp 10 \cos Z)/\sin Z$. The line crosses the sides of the square if $-10 \leqslant x \leqslant 10$.

The confidence ellipse is centred at the calculated position of the fix and is drawn by connecting up the rectangular coordinates (x, y) of a series of points that lie on the ellipse, which is specified by the previous calculated values of a, b and θ. The values of x and y are calculated from

$$x = a \cos \alpha \sin \theta - b \sin \alpha \cos \theta + 60\, d\lambda \cos \phi_f$$
$$y = a \cos \alpha \cos \theta + b \sin \alpha \sin \theta + 60\, d\phi$$

where α takes the value from 0° to 360° in steps of, say, 15°. By trial and error find the optimum step length for α below which there is no noticeable improvement in the smoothness of the ellipse which the computer draws.

At the next iteration move the origin to the latest calculated position of the fix.

Example On 1986 June 15 at $21^h\,00^m\,00^s$ GMT the dead reckoning (DR) position of a ship was W 15° 12′, N 32° 30′. At $17^h\,30^m\,45^s$ the sextant altitude of the lower limb of the Sun was 29°·9857. At $18^h\,15^m\,24^s$ the sextant altitude of the lower limb of the

Moon was 56°·9752. After sunset the sextant altitude of *Vega* (No 49) at $20^h\ 12^m\ 20^s$ was 21°·4853, and of *Dubhe* (No 27) at $20^h\ 23^m\ 15^s$ was 55°·2767. The ship maintained a constant track and speed of 315° and 12 knots, respectively. Calculate the position of the ship at $21^h\ 00^m\ 00^s$, assuming that height of eye above the horizon is 6 m and the sextant index error is zero, using the formulae and methods in section 7.

The time of fix $t_f = 21^h{\cdot}0000$. Adopting the DR position as the estimated position at the time of fix then $\lambda_f = -15°{\cdot}2000$, $\phi_f = +32°{\cdot}5000$. Intermediate values obtained in the calculation are shown in the table.

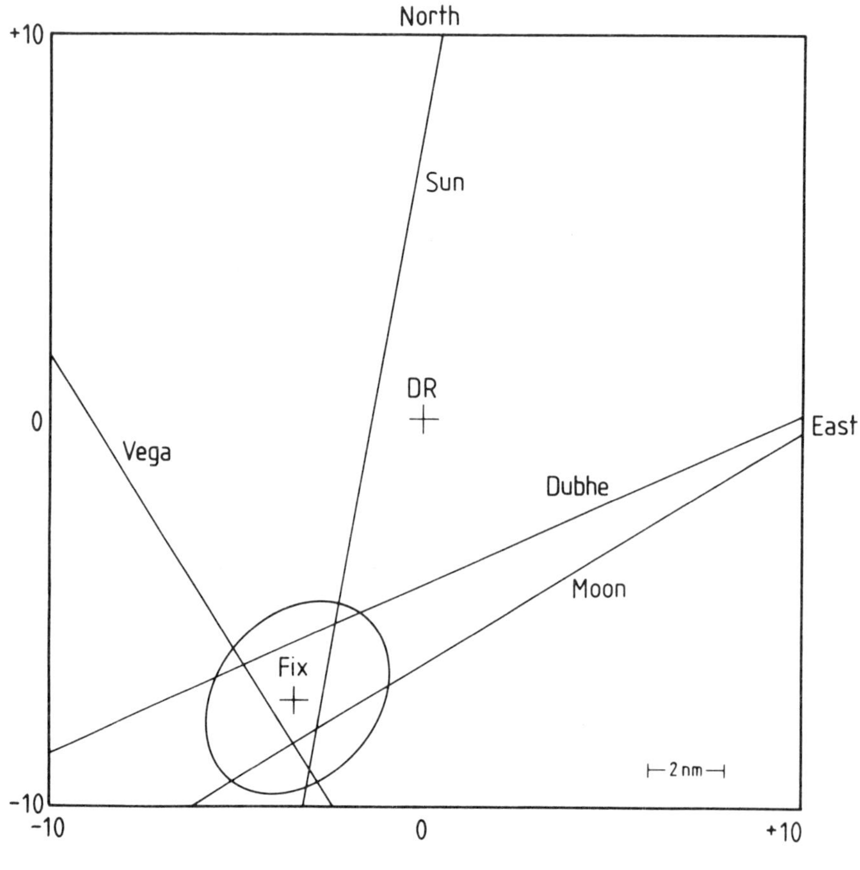

Figure: The four position lines shown correspond to the first iteration in the numerical example on page xxiii. The chart is centred at the DR position and the length of the side is 20 nautical miles (nm). The confidence ellipse is drawn for the 95% probability level.

Body	Sun	Moon	*Vega*	*Dubhe*
t	$17^h\ 30^m\ 45^s$	$18^h\ 15^m\ 24^s$	$20^h\ 12^m\ 20^s$	$20^h\ 23^m\ 15^s$
	17·5125	18·2567	20·2056	20·3875
λ (section 7.3)	−14·6152	−14·7400	−15·0668	−15·0973
ϕ	+32·0068	+32·1120	+32·3876	+32·4134
GHA (sections 2, 3, 5)	82·5829	358·7759	287·7705	43·9070
DEC	+23·3211	+3·3713	+38·7668	+61·8305
HP or *S*	0·2629	0·9605		
LHA (section 7.2)	67·9677	344·0359	272·7037	28·8097
Z	280·1973	149·1893	56·8311	336·4710
H_c	30·1285	57·5859	21·4970	55·2592
H_S (section 7.1)	29·9857	56·9752	21·4853	55·2767
D_h	0·0718	0·0718	0·0718	0·0718
I	0·0000	0·0000	0·0000	0·0000
H	29·9139	56·9034	21·4135	55·2049
R	0·0282	0·0106	0·0413	0·0113
PA	0·0021	0·5245		
Oblateness of Earth (3.3)		−0·0026		
S	0·2629	0·2617		
H_o	30·1507	57·6765	21·3722	55·1937
p (section 7.3)	+0·0222	+0·0906	−0·1248	−0·0656
	(+1·3 nm)	(+5·4 nm)	(−7·5 nm)	(−3·9 nm)

First iteration

$A = 1{\cdot}9089$	$B = -0{\cdot}5222$	$C = 2{\cdot}0911$
$D = -0{\cdot}2023$	$E = -0{\cdot}0537$	$F = 0{\cdot}0286$
$a = 2{\cdot}756$	$b = 2{\cdot}101$	$\theta = 40{\cdot}0547$
$P = 95\%$	$n = 4$	$\sigma = 1{\cdot}3651$
$\lambda_f = 15{\cdot}2664$		$\phi_f = 32{\cdot}3787$
= W 15° 16′·0		= N 32° 22′·7

The figure shows the four position lines and the confidence ellipse after the first iteration. The coordinates of the points where the position lines cross the sides of the square are

Sun	(−3·2, −10);	(0·4, 10)
Moon	(−6·2, −10);	(10, −0·4)
Vega	(−2·4, −10);	(−10, 1·6)
Dubhe	(−10, −8·6);	(10, 0·1)

Second iteration

$A = 1{\cdot}9083$	$B = -0{\cdot}5229$	$C = 2{\cdot}0917$
$D = -0{\cdot}0005$	$E = 0{\cdot}0016$	$F = 0{\cdot}0011$
$a = 2{\cdot}804$	$b = 2{\cdot}136$	$\theta = 40{\cdot}0284$
$P = 95\%$	$n = 4$	$\sigma = 1{\cdot}3883$
$\lambda_f = 15{\cdot}2655$		$\phi_f = 32{\cdot}3787$
= W 15° 15′·9		= N 32° 22′·7

8. TIMES OF RISING AND SETTING AND TRANSIT

8.1 *Times of rising and setting* The times of rising and setting of a celestial body at longitude λ, latitude ϕ may be calculated from its *GHA* and *DEC* by using the following iterative process.

Method

Step 0 Start with an initial estimate T_0 for the GMT of the phenomenon.

For the Sun a suitable estimate is $T_0 = 6^h - \lambda^h$ for sunrise and $T_0 = 18^h - \lambda^h$ for sunset, where λ^h is the longitude (in hours) measured east of the Greenwich meridian.

For any other body take $T_0 = 6^h - \lambda^h + (12^h - GHA^h$ of body at 0^h GMT) for time of rising. Replace 6^h by 18^h when estimating time of setting. If T_0 is negative add 24^h. Remove multiples of 24^h to make T_0 less than 24^h.

Calculate H the adopted altitude of the body at rising or setting. When a star or planet is on the horizon $H = -0°{\cdot}567 - 0°{\cdot}0347h^{\frac{1}{2}}$. When the upper limb of the Sun is on the horizon $H = -0°{\cdot}833 - 0°{\cdot}0347h^{\frac{1}{2}}$ where h is the height of the observer above the horizon in metres. A standard allowance of $34' = 0°{\cdot}567$ is made for refraction and $16' = 0°{\cdot}267$ for the Sun's semi-diameter. For civil, nautical and astronomical twilight $H = -6°$, $-12°$ and $-18°$, respectively.

Step 1 Calculate *GHA* (in hours) and *DEC* (in degrees) of the body at time T_0. For the Moon determine the horizontal parallax *HP* in degrees and calculate the adopted altitude for the upper limb on the horizon from

$$H = -0°{\cdot}567 + 0°{\cdot}7276\, HP - 0°{\cdot}0347h^{\frac{1}{2}}$$

Step 2 Calculate t, the local hour angle of the body in degrees at the estimated time of the phenomenon, from

$$\cos t = (\sin H - \sin\phi \sin DEC)/(\cos\phi \cos DEC)$$

If $\cos t \geqslant 1$ set $t = 0°$, if $\cos t \leqslant -1$ set $t = 180°$.

Step 3 Calculate an improved estimate T_1 of the time of the phenomenon from

$$T_1 = \pm t/15 - \lambda^h - (GHA^h - T_0)$$

where $-t$ is used for rising and $+t$ is used for setting. Add or subtract multiples of 24^h to bring T_1 as near as possible to the previous estimate T_0.

Step 4 Compare T_1 with T_0. If they differ by more than $0^h{\cdot}08$ set $T_0 = T_1$ and return to Step 1, otherwise go to Step 5.

Step 5 If the absolute value of $\cos t$ does not exceed unity, accept T_1 as the GMT of the phenomenon. If after several iterations the absolute value of $\cos t$ exceeds unity, it is probable that the phenomenon does not occur. When $\cos t \geqslant 1$ the body remains below the horizon all day. When $\cos t \leqslant -1$ the body remains above the horizon all day.

Note (1) There is always one day in each lunation around first quarter when the Moon does not set, and another around last quarter when the Moon does not rise.

(2) The algorithm may fail on rare occasions for latitudes above about 68°.

(3) It is possible, on rare occasions, for more than one moonrise or moonset to occur on the same day at high latitudes near limiting cases of no phenomenon and when the Moon is moving rapidly in declination. The extra times may usually be found by using different initial estimates for T_0, eg choose one near the beginning of the day and another near the end.

(4) The local standard time of the phenomenon is obtained by adding a correction when places are fast on GMT (mainly those east of Greenwich) and subtracting a correction when places are slow on GMT (west of Greenwich).

Example To calculate the time of moonset on 1986 April 8 for an observer at longitude W 75°, latitude N 30°, when the observer is assumed to be at the same height as the horizon. Standard Time = $GMT - 5^h$.

From Table 2 $GHA^h = ((0{\cdot}0382x + 349{\cdot}9472)\,x + 191{\cdot}7643)/15$
$DEC^\circ = (-0{\cdot}0828x + 5{\cdot}9296)\,x - 0{\cdot}7451$
$HP^\circ = -0{\cdot}0071x + 0{\cdot}9258$

At 0^h GMT on 1986 April 8, $x = 0$, $GHA^h = 12{\cdot}8$

$$\phi = +30^\circ \quad \lambda = -5^h \quad h = 0{\cdot}0$$

Step 0 $T_0 = 18^h + 5^h + (12^h - 12^h{\cdot}8) = 22^h{\cdot}2$

Step 1 $x = 22{\cdot}2/24 = 0{\cdot}925655$
$GHA = 10^h{\cdot}38$ (remove multiples of 24^h)
$DEC = +4^\circ{\cdot}67 \quad HP = 0^\circ{\cdot}92 \quad H = 0^\circ{\cdot}10$

Step 2 $t = 92^\circ{\cdot}60$

Step 3 $T_1 = 6^h{\cdot}17 + 5^h - (10^h{\cdot}38 - 22^h{\cdot}2) = 23^h{\cdot}00$

Step 4 $T_1 - T_0 = 0^h{\cdot}80$. Set $T_0 = 23^h{\cdot}00$ and return to Step 1

Step 1 $x = 23{\cdot}00/24 = 0{\cdot}958642$
$GHA = 11^h{\cdot}15 \quad DEC = +4^\circ{\cdot}86 \quad HP = 0^\circ{\cdot}92 \quad H = 0^\circ{\cdot}10$

Step 2 $t = 92^\circ{\cdot}71$

Step 3 $T_1 = 6^h{\cdot}18 + 5^h - (11^h{\cdot}15 - 23^h{\cdot}00) = 23^h{\cdot}04$

Step 4 $T_1 - T_0 = 0^h{\cdot}04$

Step 5 GMT of moonset = $23^h\,02^m$ Standard time of moonset = $18^h\,02^m$.

8.2 *Times of meridian transit* The Greenwich mean time (T) in hours of upper meridian transit may be calculated for longitude λ^h using the polynomial data in this booklet as follows:

a) The Sun Calculate $T_1 = 12 - \lambda^h$

Then $T = 24 - \lambda^h - y_1$

where y_1 is $(GHA - GMT)^h$ evaluated at time T_1 on the date required from the polynomial coefficients in Table 1.

b) The planets Use the polynomial coefficients for $(GHA - GMT)^h$ given in Table 1.

Calculate $T_1 = (24 - \lambda^h - y_0)/1{\cdot}002738$

where y_0 is $(GHA - GMT)^h$ evaluated at $GMT = 0$ for the required date. If necessary set $(24 - \lambda^h - y_0)$ in the range 0^h to 24^h.

Then $T = 24 - \lambda^h - y_1$

where y_1 is $(GHA - GMT)^h$ evaluated at time T_1.

c) The Moon Use the daily polynomial coefficients for *GHA* of the Moon given in Table 2.

Calculate $x_1 = (360 - \lambda^\circ - a_0)/a_1$

If necessary set $(360 - \lambda^\circ - a_0)$ in the range 0° to 360°.

If $x_1 > 1$ try the next day instead.

Then $T = 24\, x_1(1 - a_2 x_1/a_1)$

d) The bright stars Calculate $y_0 = GHA$ at 0^h GMT on the date required using Tables 5 and 6.

Set $(360 - \lambda^\circ - y_0)$ in the range 0° to 360°.

Then $T = (360 - \lambda^\circ - y_0)/15{\cdot}0411$

Once a year for a star there will be two transits on the same day. This occurs when T is less than about $0^h{\cdot}0655$. The second transit is obtained by adding $23^h{\cdot}9345$ to T.

The times of upper meridian transit may also be calculated from the algorithm for rising and setting but with the local hour angle t set equal to 0° in Step 2 of section 8.1. For lower transit set $t = 180°$.

9. HINTS FOR PROGRAMMERS

The following notes are intended as an elementary guide to the problem of translating algebraic expressions into a calculator code or into the programming language of a personal computer.

In mathematical formulae it is normally assumed that the operations of multiplication $(\times)$ and division $(\div)$ are carried out before addition $(+)$ and subtraction $(-)$ and terms which are enclosed in parentheses or brackets are evaluated first. For example the expression $ab + c/d$ is evaluated as $((a \times b) + (c \div d)$. Note that multiplication is implied by the term ab and that the symbol / means divide $(\div)$.

For calculators which use algebraic notation or more advanced microprocessors which use a high-level language such as BASIC, the most straightforward way to program an algebraic expression is to start at the left hand side and work to the right. With calculators that use reverse Polish notation (RPN) it is usually possible to start at the innermost pair of parentheses in the expression and work outwards, and in general the parentheses will not appear in the final calculator program. A similar technique may be used for calculators using algebraic notation, as is demonstrated in the following example:

Consider the simple polynomial expression $ax^2 + bx + c$ which may be written $((ax + b)\,x + c)$

Assume that the quantities a, b, c and x have been stored in the calculator, then with an RPN calculator starting at the innermost parentheses $(ax + b)$ the expression may be programmed as follows;

$$\text{RCL}a\ \text{RCL}x \times \text{RCL}b + \text{RCL}x \times \text{RCL}c +$$

where RCLa means recall the quantity a from the appropriate memory store.

In a calculator using algebraic notation this expression may be programmed directly as

$$((\text{RCL}a \times \text{RCL}x + \text{RCL}b) \times \text{RCL}x + \text{RCL}c)$$

Alternatively starting at the innermost pair of parentheses it may be programmed as

$$\text{RCL}a \times \text{RCL}x + \text{RCL}b = \times\ \text{RCL}x + \text{RCL}c =$$

With a little practice with most calculators, irrespective of whether they use scientific or reverse Polish notation, a polynomial of any order may be evaluated by entering values directly from the keyboard starting from the innermost pair of parentheses.

There are no hard and fast rules about programming. It is best to consult the manual to exploit the calculator to the full. Since most programs will be modified at a later stage or will be used as subroutines in other programs it is advisable to keep programs as simple and straightforward as possible and to avoid using short cuts or tricks.

Table 1: SUN AND PLANETS, 1986

		January GHA-GMT ʰ	January DEC °	February GHA-GMT ʰ	February DEC °	March GHA-GMT ʰ	March DEC °	April GHA-GMT ʰ	April DEC °
Sun	a_0	11.95346	-23.1203	11.77716	-17.5249	11.78841	-8.1594	11.92667	3.9544
	a_1	-0.25436	2.2600	-0.08226	8.8513	0.09736	12.0589	0.16103	12.4071
	a_2	0.02877	3.9862	0.11133	2.7084	0.07032	1.0684	-0.01655	-0.5755
	a_3	0.06582	-0.2694	0.00836	-0.7091	-0.01940	-0.6787	-0.02142	-0.6129
	a_4	-0.01898	-0.1024	-0.01296	0.0377	-0.00501	0.0523	-0.00028	0.0507
check sum		11.77471	-17.2459	11.80163	-6.6366	11.93168	4.3415	12.04945	15.2238
Venus	a_0	12.29335	-23.6571	11.56987	-17.9725	11.16319	-5.8127	10.84546	9.8342
	a_1	-0.82822	0.1246	-0.60618	10.9592	-0.35119	15.9304	-0.36321	15.3113
	a_2	-0.00681	6.3271	0.17868	4.4072	0.08426	1.3342	-0.09414	-1.8891
	a_3	0.12372	-0.1265	-0.00278	-1.2203	-0.06228	-1.1577	-0.05798	-1.2482
	a_4	-0.03093	-0.2940	-0.01912	0.0235	0.00004	0.0165	0.02200	0.0725
check sum		11.55111	-17.6259	11.12047	-3.8029	10.83402	10.3107	10.35213	22.0807
Mars	a_0	16.09855	-13.8221	16.88715	-19.0627	17.59182	-22.0929	18.43429	-23.5438
	a_1	0.83096	-6.2990	0.80079	-4.4388	0.81898	-2.4874	0.94033	-0.6350
	a_2	-0.02205	0.8052	-0.00542	1.0892	0.03081	1.1142	0.10709	0.7384
	a_3	0.00534	0.1452	0.01167	0.0567	0.02241	-0.0568	0.02540	-0.2488
	a_4	-0.00062	-0.0297	0.00052	-0.0309	-0.00025	-0.0401	0.00595	-0.0375
check sum		16.91218	-19.2004	17.69471	-22.3865	18.46377	-23.5630	19.51306	-23.7267
Jupiter	a_0	9.24176	-16.1893	10.82027	-13.9344	12.22983	-11.6319	13.81053	-9.0317
	a_1	1.65500	2.0930	1.61292	2.5239	1.61499	2.7000	1.65486	2.6145
	a_2	-0.03687	0.2822	-0.01024	0.1687	0.00953	0.0431	0.02996	-0.1220
	a_3	0.01207	-0.0408	0.01039	-0.0544	0.00897	-0.0659	0.00832	-0.0724
	a_4	-0.00128	-0.0005	-0.00147	0.0024	-0.00106	0.0046	-0.00063	0.0060
check sum		10.87068	-13.8554	12.43187	-11.2938	13.86226	-8.9501	15.50304	-6.6056
Saturn	a_0	14.39366	-19.3682	16.23349	-19.8016	17.96693	-19.9746	19.98882	-19.9333
	a_1	1.86821	-0.5727	1.93677	-0.3177	2.02894	-0.0803	2.14482	0.1570
	a_2	0.02419	0.1202	0.04549	0.1385	0.05855	0.1339	0.05891	0.1151
	a_3	0.00894	0.0142	0.00702	-0.0003	0.00350	-0.0062	-0.00517	-0.0172
	a_4	-0.00095	-0.0050	-0.00129	-0.0014	-0.00202	-0.0011	-0.00165	-0.0006
check sum		16.29405	-19.8115	18.22148	-19.9825	20.05590	-19.9283	22.18573	-19.6790

Semi-diameter (S) of the Sun and HP of Venus and Mars

		January °	February °	March °	April °
Sun S	a_0	0.2716	0.2710	0.2695	0.2672
	a_1	-0.0006	-0.0018	-0.0024	-0.0023
Venus HP	a_0	0.0014	0.0014	0.0014	0.0015
	a_1	0.0000	0.0000	0.0001	0.0001
Mars HP	a_0	0.0013	0.0015	0.0018	0.0023
	a_1	0.0002	0.0004	0.0006	0.0009

$$GHA^\circ = 15((((a_4 x + a_3)x + a_2)x + a_1)x + a_0 + GMT^h)$$

$$DEC^\circ = (((a_4 x + a_3)x + a_2)x + a_1)x + a_0$$

$$S^\circ,\ HP^\circ = a_1 x + a_0 \qquad \text{where } x = (d + GMT^h/24)/32$$

		May GHA-GMT (h)	May DEC (°)	June GHA-GMT (h)	June DEC (°)	July GHA-GMT (h)	July DEC (°)	August GHA-GMT (h)	August DEC (°)
Sun	a_0	12.04522	14.6144	12.04156	21.8428	11.94255	23.2059	11.89380	18.3969
	a_1	0.07299	9.8768	-0.07472	4.6581	-0.10639	-1.8160	0.02312	-7.8109
	a_2	-0.08099	-2.0153	-0.06814	-3.2368	0.02602	-3.5067	0.08545	-2.6053
	a_3	-0.00192	-0.5727	0.04108	-0.3070	0.04726	0.2099	0.00489	0.5111
	a_4	0.00386	0.0820	-0.00378	0.1217	-0.01483	0.0571	-0.01077	-0.0219
check sum		12.03916	21.9852	11.93600	23.0788	11.89461	18.1502	11.99649	8.4699
Venus	a_0	10.39139	21.5557	9.71755	24.7060	9.20955	17.6140	9.06263	3.7308
	a_1	-0.61855	8.7136	-0.69146	-2.5279	-0.35060	-11.8265	0.02345	-15.8437
	a_2	-0.15164	-4.9527	0.08637	-6.0504	0.22095	-3.6690	0.15105	-0.6541
	a_3	0.06227	-1.0631	0.10862	0.5663	-0.00658	1.1836	-0.02799	1.0468
	a_4	0.01258	0.3675	-0.03258	0.1669	-0.00981	-0.0669	0.01823	-0.0622
check sum		9.69605	24.6210	9.18850	16.8609	9.06351	3.2352	9.22737	-11.7824
Mars	a_0	19.43550	-23.7241	20.88467	-24.3041	22.98771	-26.7700	1.60996	-28.7075
	a_1	1.22687	-0.0327	1.82656	-1.5397	2.63435	-3.2163	2.50863	-0.2490
	a_2	0.22503	-0.1465	0.40944	-1.3182	0.39762	0.1147	-0.48284	1.9196
	a_3	0.04475	-0.5001	0.09476	-0.2303	-0.36989	1.7783	-0.09284	-0.6673
	a_4	0.00998	0.0497	-0.06145	0.4221	0.03801	-0.6201	0.06305	0.1763
check sum		20.94213	-24.3537	23.15398	-26.9702	25.68780	-28.7134	3.60596	-27.5279
Jupiter	a_0	15.39466	-6.7429	17.12707	-4.9416	18.93866	-4.0697	20.98856	-4.3199
	a_1	1.73085	2.2143	1.85334	1.4456	2.01792	0.3727	2.21323	-0.8716
	a_2	0.05113	-0.3002	0.07676	-0.4983	0.09934	-0.6491	0.09779	-0.6077
	a_3	0.00872	-0.0736	0.00857	-0.0609	0.00478	-0.0218	-0.01280	0.1079
	a_4	-0.00030	0.0056	-0.00056	0.0063	-0.00289	0.0201	-0.00421	0.0234
check sum		17.18506	-4.8968	19.06518	-4.0489	21.05781	-4.3478	23.28257	-5.6679
Saturn	a_0	22.04583	-19.6996	0.23412	-19.3499	2.34614	-19.0562	4.45879	-18.9627
	a_1	2.23611	0.3253	2.26849	0.3681	2.22616	0.2308	2.13022	-0.0506
	a_2	0.03714	0.0671	-0.00312	-0.0245	-0.03927	-0.1192	-0.05730	-0.1643
	a_3	-0.01414	-0.0313	-0.01643	-0.0403	-0.01000	-0.0268	-0.00091	0.0035
	a_4	0.00006	0.0000	0.00206	0.0044	0.00226	0.0070	0.00098	0.0034
check sum		24.30500	-19.3385	2.48512	-19.0422	4.52529	-18.9644	6.53178	-19.1707

Semi-diameter (S) of the Sun and HP of Venus and Mars

		May (°)	June (°)	July (°)	August (°)
Sun S	a_0	0.2649	0.2632	0.2625	0.2630
	a_1	-0.0017	-0.0007	0.0005	0.0016
Venus HP	a_0	0.0016	0.0018	0.0021	0.0027
	a_1	0.0002	0.0004	0.0006	0.0011
Mars HP	a_0	0.0031	0.0045	0.0059	0.0058
	a_1	0.0013	0.0014	0.0000	-0.0013

$$GHA^{\circ} = 15((((a_4x + a_3)x + a_2)x + a_1)x + a_0 + GMT^{h})$$

$$DEC^{\circ} = (((a_4x + a_3)x + a_2)x + a_1)x + a_0$$

$$S^{\circ},\ HP^{\circ} = a_1x + a_0 \qquad \text{where } x = (d + GMT^{h}/24)/32$$

Table 1: SUN AND PLANETS, 1986

		September GHA-GMT (h)	September DEC (°)	October GHA-GMT (h)	October DEC (°)	November GHA-GMT (h)	November DEC (°)	December GHA-GMT (h)	December DEC (°)
Sun	a_0	11.99136	8.8305	12.16295	-2.5964	12.27231	-13.9398	12.19319	-21.5607
	a_1	0.16325	-11.4998	0.17462	-12.4476	0.02388	-10.4440	-0.19102	-5.3177
	a_2	0.04721	-1.2097	-0.03452	0.2245	-0.10729	1.9052	-0.09544	3.5129
	a_3	-0.02411	0.5217	-0.03168	0.5488	-0.01956	0.6370	0.03625	0.4541
	a_4	-0.00398	-0.0161	0.00157	0.0064	0.01154	-0.0376	0.00448	-0.1534
check sum		12.17373	-3.3734	12.27294	-14.2643	12.18088	-21.8792	11.94746	-23.0648
Venus	a_0	9.21773	-11.3346	9.77302	-22.1912	11.83184	-22.3666	14.36139	-12.5956
	a_1	0.30342	-14.3967	1.09878	-7.5772	3.18350	9.1664	1.64996	4.1010
	a_2	0.18401	2.1001	0.51449	3.8645	0.70009	14.2912	-1.15553	-10.0015
	a_3	0.05899	0.8371	1.06448	4.9954	-1.96081	-18.8675	0.37573	3.4331
	a_4	0.07924	0.1416	-0.51817	-1.1502	0.70571	5.3942	-0.06750	-0.1703
check sum		9.84339	-22.6525	11.93260	-22.0587	14.46033	-12.3823	15.16405	-15.2333
Mars	a_0	3.55814	-27.5987	4.70059	-24.5589	5.52741	-19.1159	6.23340	-11.7368
	a_1	1.54023	2.2350	0.97928	4.3409	0.77535	6.8595	0.74580	8.7168
	a_2	-0.44412	0.9795	-0.19527	1.3229	-0.05043	1.2640	-0.00163	0.7572
	a_3	0.12274	0.0935	0.08449	0.0291	0.03980	-0.2071	0.01257	-0.2850
	a_4	-0.01594	0.0084	-0.01751	-0.0344	-0.01212	0.0103	-0.00697	0.0330
check sum		4.76105	-24.2823	5.55158	-18.9004	6.28001	-11.1892	6.98317	-2.5148
Jupiter	a_0	23.20905	-5.6159	1.41754	-7.1299	3.60232	-7.9634	5.53525	-7.6025
	a_1	2.35125	-1.6603	2.33163	-1.3913	2.16317	-0.2518	1.96338	0.9969
	a_2	0.03689	-0.1619	-0.05697	0.4329	-0.10915	0.6982	-0.10121	0.6273
	a_3	-0.03576	0.2430	-0.02825	0.1546	-0.00117	-0.0120	0.01178	-0.0727
	a_4	0.00161	-0.0201	0.00586	-0.0369	0.00239	-0.0088	-0.00061	0.0062
check sum		25.56304	-7.2152	3.66981	-7.9706	5.65756	-7.5378	7.40859	-6.0448
Saturn	a_0	6.46870	-19.1599	8.31765	-19.5868	10.15120	-20.1546	11.88197	-20.6997
	a_1	2.02026	-0.3470	1.92824	-0.5427	1.86286	-0.6056	1.83543	-0.5403
	a_2	-0.05558	-0.1384	-0.04294	-0.0725	-0.02436	0.0057	-0.00401	0.0619
	a_3	0.00468	0.0255	0.00677	0.0316	0.00708	0.0232	0.00710	0.0149
	a_4	-0.00005	-0.0012	-0.00033	-0.0032	-0.00012	-0.0019	0.00011	-0.0027
check sum		8.43801	-19.6210	10.20939	-20.1736	11.99666	-20.7332	13.72060	-21.1659

Semi-diameter (S) of the Sun and HP of Venus and Mars

		September (°)	October (°)	November (°)	December (°)
Sun S	a_0	0.2645	0.2667	0.2690	0.2708
	a_1	0.0023	0.0024	0.0019	0.0008
Venus HP	a_0	0.0037	0.0058	0.0093	0.0069
	a_1	0.0022	0.0033	-0.0023	-0.0027
Mars HP	a_0	0.0046	0.0035	0.0027	0.0022
	a_1	-0.0011	-0.0008	-0.0006	-0.0004

$$GHA^{\circ} = 15((((a_4x + a_3)x + a_2)x + a_1)x + a_0 + GMT^{h})$$

$$DEC^{\circ} = (((a_4x + a_3)x + a_2)x + a_1)x + a_0$$

$$S^{\circ},\ HP^{\circ} = a_1x + a_0 \qquad \text{where } x = (d + GMT^{h}/24)/32$$

Table 1: SUN AND PLANETS, 1987

		January		February		March		April	
		GHA-GMT	DEC	GHA-GMT	DEC	GHA-GMT	DEC	GHA-GMT	DEC
		h	°	h	°	h	°	h	°
Sun	a_0	11.95548	-23.1377	11.77776	-17.5917	11.78763	-8.2500	11.92521	3.8628
	a_1	-0.25802	2.2064	-0.08629	8.8273	0.09363	12.0648	0.15891	12.4347
	a_2	0.04225	3.9760	0.12445	2.6447	0.08240	0.9938	-0.00377	-0.6637
	a_3	0.04705	-0.2633	-0.01072	-0.6056	-0.03392	-0.5729	-0.03885	-0.4664
	a_4	-0.01159	-0.0946	-0.00464	-0.0022	0.00041	0.0151	0.00696	-0.0134
check sum		11.77517	-17.3132	11.80056	-6.7275	11.93015	4.2508	12.04846	15.1540
Venus	a_0	15.15753	-15.0377	15.04725	-20.3458	14.62684	-19.7626	14.18956	-11.1564
	a_1	0.22222	-6.2457	-0.37635	-3.1384	-0.52083	4.7728	-0.36156	12.3345
	a_2	-0.42043	-1.1005	-0.19149	3.7093	0.03137	4.7552	0.10317	2.6895
	a_3	0.07441	2.2295	0.08194	1.0655	0.06086	-0.3468	-0.03451	-0.9074
	a_4	0.00155	-0.2858	0.00102	-0.3830	-0.01988	-0.1870	-0.00514	-0.0406
check sum		15.03528	-20.4402	14.56237	-19.0924	14.17836	-10.7684	13.89152	2.9196
Mars	a_0	6.95965	-2.8119	7.68407	6.2750	8.30992	13.6804	8.94505	20.0657
	a_1	0.75269	9.5011	0.73576	9.0699	0.68935	7.7222	0.62074	5.3243
	a_2	-0.00087	0.0984	-0.01799	-0.5253	-0.03368	-1.0074	-0.03236	-1.4552
	a_3	-0.00115	-0.2599	-0.00574	-0.2066	-0.00122	-0.1849	0.00436	-0.1162
	a_4	-0.00329	0.0303	-0.00055	0.0165	0.00004	0.0205	0.00123	0.0190
check sum		7.70703	6.5580	8.39555	14.6295	8.96441	20.2308	9.53902	23.8376
Jupiter	a_0	7.35246	-6.1087	9.03914	-3.7864	10.49644	-1.2637	12.07828	1.7002
	a_1	1.79819	2.0300	1.69326	2.7069	1.64350	3.0168	1.62803	3.0553
	a_2	-0.06970	0.4449	-0.03892	0.2515	-0.01725	0.0956	0.00294	-0.0701
	a_3	0.01227	-0.0749	0.00866	-0.0585	0.00665	-0.0556	0.00452	-0.0350
	a_4	-0.00120	0.0071	-0.00052	0.0013	-0.00018	0.0025	0.00079	-0.0055
check sum		9.09202	-3.7016	10.70162	-0.8852	12.12916	1.7956	13.71456	4.6449
Saturn	a_0	13.66283	-21.1539	15.47669	-21.4437	17.17670	-21.5636	19.15603	-21.5705
	a_1	1.84803	-0.3888	1.90351	-0.2099	1.98673	-0.0685	2.10174	0.0488
	a_2	0.01841	0.0935	0.03986	0.0885	0.05544	0.0702	0.06099	0.0485
	a_3	0.00664	-0.0003	0.00584	-0.0058	0.00359	-0.0079	-0.00128	-0.0031
	a_4	0.00030	-0.0005	0.00001	-0.0003	-0.00069	0.0008	-0.00170	-0.0004
check sum		15.53621	-21.4500	17.42591	-21.5712	19.22177	-21.5690	21.31578	-21.4767

Semi-diameter (S) of the Sun and HP of Venus and Mars

		January	February	March	April
		°	°	°	°
Sun S	a_0	0.2716	0.2710	0.2695	0.2672
	a_1	-0.0006	-0.0018	-0.0024	-0.0024
Venus HP	a_0	0.0043	0.0030	0.0024	0.0020
	a_1	-0.0013	-0.0007	-0.0004	-0.0003
Mars HP	a_0	0.0018	0.0015	0.0013	0.0012
	a_1	-0.0003	-0.0002	-0.0002	-0.0001

$$GHA^{\circ} = 15((((a_4 x + a_3)x + a_2)x + a_1)x + a_0 + GMT^{h})$$

$$DEC^{\circ} = (((a_4 x + a_3)x + a_2)x + a_1)x + a_0$$

$$S^{\circ},\ HP^{\circ} = a_1 x + a_0 \qquad \text{where } x = (d + GMT^{h}/24)/32$$

Table 1: SUN AND PLANETS, 1987

		May		June		July		August	
		GHA-GMT	DEC	GHA-GMT	DEC	GHA-GMT	DEC	GHA-GMT	DEC
		h	°	h	°	h	°	h	°
Sun	a_0	12.04425	14.5424	12.04153	21.8092	11.94277	23.2192	11.89321	18.4539
	a_1	0.07169	9.9183	-0.07567	4.7038	-0.10810	-1.7741	0.02183	-7.7808
	a_2	-0.06704	-2.0793	-0.05428	-3.2421	0.03839	-3.4637	0.09399	-2.5387
	a_3	-0.02241	-0.4529	0.01690	-0.2774	0.02436	0.1427	-0.01276	0.3758
	a_4	0.01261	0.0245	0.00770	0.1013	-0.00343	0.0843	-0.00119	0.0453
check sum		12.03910	21.9530	11.93618	23.0948	11.89399	18.2084	11.99508	8.5555
Venus	a_0	13.90887	1.9920	13.57558	15.3792	13.01722	22.8741	12.31980	20.7650
	a_1	-0.27675	14.8534	-0.45597	11.7971	-0.71431	3.3952	-0.65445	-7.5944
	a_2	-0.01323	-0.0979	-0.14380	-3.0775	-0.08604	-5.5863	0.13575	-5.1413
	a_3	-0.06878	-0.9560	-0.03374	-1.1792	0.08863	-0.5062	0.02923	0.8406
	a_4	0.01106	-0.0466	0.03018	0.1446	-0.00599	0.3459	-0.02051	0.0945
check sum		13.56117	15.7449	12.97225	23.0642	12.29951	20.5227	11.80982	8.9644
Mars	a_0	9.50310	23.6972	10.06121	24.3826	10.64018	22.1340	11.31873	17.2994
	a_1	0.57540	2.3530	0.58672	-0.9251	0.65406	-3.7882	0.74584	-6.0678
	a_2	-0.00954	-1.6925	0.02617	-1.6493	0.04865	-1.3628	0.04831	-0.9533
	a_3	0.01047	-0.0259	0.00761	0.0685	-0.00070	0.1160	-0.00849	0.1177
	a_4	0.00014	0.0203	-0.00046	0.0152	-0.00010	0.0098	0.00106	0.0109
check sum		10.07957	24.3521	10.68125	21.8919	11.34209	17.1088	12.10545	10.4069
Jupiter	a_0	13.61147	4.4698	15.23319	6.9693	16.87747	8.8069	18.70369	9.8767
	a_1	1.64797	2.8145	1.70690	2.3048	1.80936	1.5754	1.97106	0.5923
	a_2	0.02120	-0.2071	0.04259	-0.3319	0.06828	-0.4460	0.09571	-0.5453
	a_3	0.00449	-0.0286	0.00655	-0.0319	0.00988	-0.0409	0.01152	-0.0496
	a_4	0.00144	-0.0076	0.00159	-0.0065	0.00040	-0.0007	-0.00383	0.0199
check sum		15.28657	7.0410	16.99082	8.9038	18.76539	9.8947	20.77815	9.8940
Saturn	a_0	21.17764	-21.4850	23.35080	-21.3328	1.47636	-21.1762	3.62449	-21.0978
	a_1	2.20715	0.1298	2.26831	0.1755	2.25395	0.1420	2.17233	0.0048
	a_2	0.04722	0.0340	0.01186	0.0059	-0.02784	-0.0459	-0.05365	-0.0898
	a_3	-0.00871	-0.0025	-0.01368	-0.0127	-0.01125	-0.0188	-0.00441	-0.0092
	a_4	-0.00161	-0.0035	-0.00017	-0.0035	0.00110	0.0012	0.00127	0.0049
check sum		23.42169	-21.3272	25.61712	-21.1676	3.69232	-21.0977	5.74003	-21.1871

Semi-diameter (S) of the Sun and HP of Venus and Mars

		May	June	July	August
		°	°	°	°
Sun S	a_0	0.2650	0.2632	0.2625	0.2630
	a_1	-0.0018	-0.0007	0.0004	0.0016
Venus HP	a_0	0.0017	0.0016	0.0015	0.0014
	a_1	-0.0002	-0.0001	-0.0001	0.0000
Mars HP	a_0	0.0011	0.0010	0.0009	0.0009
	a_1	-0.0001	-0.0001	0.0000	0.0000

$$GHA^\circ = 15((((a_4x + a_3)x + a_2)x + a_1)x + a_0 + GMT^h)$$

$$DEC^\circ = (((a_4x + a_3)x + a_2)x + a_1)x + a_0$$

$$S^\circ,\ HP^\circ = a_1x + a_0 \qquad \text{where } x = (d + GMT^h/24)/32$$

		September		October		November		December	
		GHA-GMT	DEC	GHA-GMT	DEC	GHA-GMT	DEC	GHA-GMT	DEC
		h	°	h	°	h	°	h	°
Sun	a_0	11.98993	8.9154	12.16155	-2.5037	12.27206	-13.8617	12.19449	-21.5209
	a_1	0.16339	-11.4796	0.17605	-12.4425	0.02709	-10.4573	-0.18760	-5.3581
	a_2	0.05218	-1.1674	-0.02997	0.2361	-0.10615	1.8608	-0.09694	3.4550
	a_3	-0.03655	0.3972	-0.04379	0.4623	-0.02865	0.6435	0.02873	0.5194
	a_4	0.00349	0.0542	0.00896	0.0611	0.01804	-0.0273	0.01069	-0.1781
check sum		12.17244	-3.2802	12.27280	-14.1867	12.18239	-21.8420	11.94937	-23.0827
Venus	a_0	11.82171	9.4304	11.50851	-5.5131	11.07128	-19.1294	10.38031	-24.7135
	a_1	-0.38407	-14.8499	-0.33935	-16.0550	-0.60189	-10.9231	-0.81451	-0.3196
	a_2	0.11092	-2.1708	-0.05609	0.9347	-0.17615	4.3074	-0.00336	6.4655
	a_3	-0.06086	1.0445	-0.08102	1.1337	0.00319	1.3424	0.11920	-0.0320
	a_4	-0.00063	0.0331	0.02023	0.0332	0.03300	-0.3052	-0.02361	-0.3324
check sum		11.48707	-6.5127	11.05228	-19.4665	10.32943	-24.7079	9.65803	-18.9320
Mars	a_0	12.07982	10.6431	12.86565	3.1951	13.68300	-4.8037	14.43264	-12.1068
	a_1	0.81928	-7.5447	0.84939	-8.2235	0.82962	-8.1519	0.76349	-7.2819
	a_2	0.02994	-0.5407	0.00468	-0.1595	-0.02346	0.2419	-0.04605	0.6772
	a_3	-0.01192	0.1072	-0.01278	0.1125	-0.01115	0.1431	-0.00749	0.1748
	a_4	0.00164	0.0158	0.00196	0.0172	0.00216	0.0114	0.00286	0.0000
check sum		12.91876	2.6807	13.70890	-5.0582	14.48017	-12.5592	15.14545	-18.5367
Jupiter	a_0	20.71007	9.9112	22.83332	8.9664	1.12486	7.5018	3.27671	6.5425
	a_1	2.17508	-0.5313	2.33977	-1.4004	2.35820	-1.4306	2.21197	-0.4944
	a_2	0.10424	-0.5554	0.05805	-0.2952	-0.04033	0.2741	-0.10361	0.6438
	a_3	-0.00161	0.0063	-0.03004	0.1664	-0.03385	0.2049	-0.00959	0.0557
	a_4	-0.00800	0.0469	-0.00260	0.0202	0.00567	-0.0360	0.00463	-0.0326
check sum		22.97978	8.8777	25.19850	7.4574	3.41455	6.5142	5.38011	6.7150
Saturn	a_0	5.67570	-21.1815	7.55812	-21.4152	9.41379	-21.7390	11.15251	-22.0347
	a_1	2.06054	-0.1779	1.95791	-0.3086	1.87851	-0.3410	1.83696	-0.2755
	a_2	-0.05801	-0.0860	-0.04838	-0.0439	-0.03130	0.0107	-0.01159	0.0543
	a_3	0.00097	0.0061	0.00388	0.0165	0.00539	0.0194	0.00650	0.0099
	a_4	0.00110	0.0045	0.00094	0.0015	0.00090	-0.0019	0.00065	-0.0020
check sum		7.68030	-21.4348	9.47247	-21.7497	11.26729	-22.0518	12.98503	-22.2480

Semi-diameter (S) of the Sun and HP of Venus and Mars

		September	October	November	December
		°	°	°	°
Sun S	a_0	0.2645	0.2667	0.2690	0.2708
	a_1	0.0023	0.0024	0.0019	0.0008
Venus HP	a_0	0.0014	0.0014	0.0015	0.0016
	a_1	0.0000	0.0001	0.0001	0.0002
Mars HP	a_0	0.0009	0.0009	0.0010	0.0010
	a_1	0.0000	0.0000	0.0001	0.0001

$$GHA^\circ = 15((((a_4x + a_3)x + a_2)x + a_1)x + a_0 + GMT^h)$$

$$DEC^\circ = (((a_4x + a_3)x + a_2)x + a_1)x + a_0$$

$$S^\circ,\ HP^\circ = a_1x + a_0 \qquad \text{where } x = (d + GMT^h/24)/32$$

Table 1: SUN AND PLANETS, 1988

		January		February		March		April	
		GHA-GMT h	DEC °	GHA-GMT h	DEC °	GHA-GMT h	DEC °	GHA-GMT h	DEC °
Sun	a_0	11.95730	-23.1537	11.77840	-17.6574	11.78995	-7.9631	11.92914	4.1565
	a_1	-0.25586	2.1469	-0.08529	8.7717	0.09963	12.0930	0.16103	12.3815
	a_2	0.03533	3.9546	0.11591	2.6933	0.07509	0.9917	-0.01421	-0.6185
	a_3	0.04848	-0.1939	-0.00325	-0.6143	-0.03256	-0.5780	-0.03176	-0.5519
	a_4	-0.00938	-0.1348	-0.00585	-0.0145	0.00206	-0.0009	0.00607	0.0123
check sum		11.77587	-17.3809	11.79992	-6.8212	11.93417	4.5427	12.05027	15.3799
Venus	a_0	9.67569	-19.2772	9.30100	-5.4981	9.15271	9.4952	9.02351	22.2865
	a_1	-0.57252	10.9115	-0.23110	16.4171	-0.13411	15.7904	-0.10773	9.8129
	a_2	0.21144	4.5594	0.12542	1.1074	0.01272	-1.7582	0.10066	-4.1864
	a_3	-0.00912	-1.1599	-0.06771	-1.0666	-0.05190	-1.0218	0.07683	-0.5331
	a_4	-0.01160	-0.0180	0.01262	-0.0005	0.04082	0.0833	0.07527	0.2946
check sum		9.29389	-4.9842	9.14023	10.9593	9.02024	22.5889	9.16854	27.6745
Mars	a_0	15.12475	-18.3668	15.71902	-22.4397	16.21287	-23.6348	16.74036	-21.9088
	a_1	0.66375	-5.4780	0.57025	-2.7982	0.53300	0.1951	0.57124	3.2792
	a_2	-0.05682	1.1550	-0.04282	1.5589	-0.00710	1.6942	0.03287	1.4672
	a_3	0.00294	0.1878	0.01742	0.1178	0.02338	-0.0239	0.02126	-0.2001
	a_4	0.00219	-0.0236	-0.00107	-0.0405	-0.00391	-0.0355	-0.00595	-0.0051
check sum		15.73681	-22.5256	16.26280	-23.6017	16.75824	-21.8049	17.35978	-17.3676
Jupiter	a_0	5.31769	6.6897	7.16541	7.9777	8.76092	9.8785	10.37501	12.2507
	a_1	2.00103	0.7922	1.82236	1.7978	1.70688	2.3310	1.63332	2.5026
	a_2	-0.10403	0.6120	-0.07656	0.3986	-0.04962	0.1857	-0.02661	0.0017
	a_3	0.00573	-0.0442	0.00856	-0.0704	0.00729	-0.0652	0.00688	-0.0567
	a_4	0.00187	-0.0154	0.00074	-0.0051	0.00056	-0.0011	0.00028	0.0026
check sum		7.22229	8.0343	8.92051	10.0986	10.42603	12.3289	11.98888	14.7009
Saturn	a_0	12.92767	-22.2433	14.72194	-22.3267	16.45398	-22.3184	18.39563	-22.2795
	a_1	1.83520	-0.1501	1.87655	-0.0283	1.95168	0.0352	2.06068	0.0341
	a_2	0.00991	0.0687	0.03126	0.0526	0.04883	0.0180	0.05975	-0.0147
	a_3	0.00780	0.0004	0.00852	-0.0111	0.00732	-0.0144	0.00314	-0.0082
	a_4	0.00004	-0.0032	-0.00089	-0.0014	-0.00171	0.0011	-0.00268	0.0033
check sum		14.78062	-22.3275	16.63738	-22.3149	18.46010	-22.2785	20.51652	-22.2650

Semi-diameter (S) of the Sun and HP of Venus and Mars

		January °	February °	March °	April °
Sun S	a_0	0.2716	0.2711	0.2695	0.2672
	a_1	-0.0006	-0.0018	-0.0024	-0.0023
Venus HP	a_0	0.0018	0.0021	0.0025	0.0033
	a_1	0.0003	0.0005	0.0008	0.0017
Mars HP	a_0	0.0011	0.0013	0.0015	0.0018
	a_1	0.0002	0.0002	0.0003	0.0004

$$GHA^\circ = 15((((a_4x + a_3)x + a_2)x + a_1)x + a_0 + GMT^h)$$

$$DEC^\circ = (((a_4x + a_3)x + a_2)x + a_1)x + a_0$$

$$S^\circ,\ HP^\circ = a_1x + a_0 \qquad \text{where } x = (d + GMT^h/24)/32$$

Table 1: SUN AND PLANETS, 1988

		May GHA-GMT (h)	May DEC (°)	June GHA-GMT (h)	June DEC (°)	July GHA-GMT (h)	July DEC (°)	August GHA-GMT (h)	August DEC (°)
Sun	a_0	12.04612	14.7753	12.03998	21.9179	11.94049	23.1749	11.89401	18.2685
	a_1	0.07115	9.8028	-0.07521	4.5468	-0.10354	-1.9287	0.02800	-7.8861
	a_2	-0.07839	-2.0363	-0.06744	-3.2305	0.02557	-3.4849	0.08042	-2.5951
	a_3	-0.01160	-0.5501	0.03624	-0.3260	0.04406	0.1788	0.00695	0.5001
	a_4	0.01029	0.0650	0.00057	0.1325	-0.01160	0.0795	-0.01013	-0.0052
check sum		12.03757	22.0567	11.93414	23.0407	11.89498	18.0196	11.99925	8.2822
Venus	a_0	9.13203	27.5948	10.65507	25.4637	13.65096	18.5359	14.96199	18.7573
	a_1	0.55012	1.5403	2.83032	-5.6779	2.49256	-4.5041	0.51270	2.3902
	a_2	0.58262	-4.2553	1.92627	-7.0487	-1.52131	7.5337	-0.60543	-0.3577
	a_3	0.62616	0.9656	-2.03358	5.6650	0.39066	-2.6889	0.20991	-2.0640
	a_4	-0.14463	-0.5675	0.42185	-0.1102	-0.03533	-0.0461	-0.03363	0.4243
check sum		10.74630	25.2779	13.79993	18.2919	14.97754	18.8305	15.04554	19.1501
Mars	a_0	17.31769	-17.7138	18.03197	-11.8122	18.88327	-5.9665	20.04922	-1.5038
	a_1	0.66967	5.4829	0.81283	6.4384	1.02072	5.7533	1.43617	3.1531
	a_2	0.05773	0.9207	0.08258	0.1189	0.14218	-0.8152	0.29531	-1.8546
	a_3	0.01671	-0.3233	0.02052	-0.3960	0.04419	-0.4343	0.10828	-0.4275
	a_4	-0.00436	0.0227	-0.00027	0.0407	0.00402	0.0558	-0.02568	0.2025
check sum		18.05744	-11.6108	18.94763	-5.6102	20.09438	-1.4069	21.86330	-0.4303
Jupiter	a_0	11.88874	14.5536	13.44110	16.6840	14.96872	18.3408	16.61890	19.5421
	a_1	1.60266	2.3640	1.60950	2.0011	1.65740	1.5173	1.75972	0.9582
	a_2	-0.00758	-0.1285	0.01264	-0.2247	0.03632	-0.2746	0.06784	-0.2925
	a_3	0.00754	-0.0509	0.00953	-0.0343	0.01213	-0.0199	0.01460	-0.0038
	a_4	0.00006	0.0080	-0.00032	0.0085	-0.00061	0.0081	-0.00143	0.0036
check sum		13.49142	16.7462	15.07245	18.4346	16.67396	19.5717	18.45963	20.2076
Saturn	a_0	20.38055	-22.2646	22.52950	-22.2830	0.65634	-22.3102	2.83041	-22.3429
	a_1	2.17208	-0.0040	2.25507	-0.0288	2.26898	-0.0286	2.20804	-0.0444
	a_2	0.05494	-0.0197	0.02829	-0.0022	-0.01171	0.0011	-0.04637	-0.0200
	a_3	-0.00508	0.0026	-0.01462	0.0031	-0.01645	-0.0066	-0.00980	-0.0055
	a_4	-0.00248	0.0018	-0.00015	-0.0010	0.00220	-0.0001	0.00279	0.0019
check sum		22.60001	-22.2839	24.79809	-22.3119	2.89936	-22.3444	4.98507	-22.4109

Semi-diameter (S) of the Sun and HP of Venus and Mars

		May (°)	June (°)	July (°)	August (°)
Sun S	a_0	0.2649	0.2632	0.2625	0.2630
	a_1	-0.0017	-0.0007	0.0005	0.0016
Venus HP	a_0	0.0048	0.0083	0.0072	0.0044
	a_1	0.0031	-0.0009	-0.0028	-0.0014
Mars HP	a_0	0.0021	0.0027	0.0034	0.0045
	a_1	0.0006	0.0008	0.0011	0.0013

$$GHA^\circ = 15((((a_4x + a_3)x + a_2)x + a_1)x + a_0 + GMT^h)$$

$$DEC^\circ = (((a_4x + a_3)x + a_2)x + a_1)x + a_0$$

$$S^\circ,\ HP^\circ = a_1x + a_0 \qquad \text{where } x = (d + GMT^h/24)/32$$

		September		October		November		December	
		GHA-GMT h	DEC °	GHA-GMT h	DEC °	GHA-GMT h	DEC °	GHA-GMT h	DEC °
Sun	a_0	11.99399	8.6434	12.16581	-2.7978	12.27270	-14.1082	12.18998	-21.6458
	a_1	0.16679	-11.5245	0.17545	-12.4254	0.02239	-10.3713	-0.19184	-5.2001
	a_2	0.03899	-1.2306	-0.04369	0.1922	-0.11838	1.8782	-0.10568	3.5118
	a_3	-0.01761	0.5609	-0.02376	0.6153	-0.00603	0.7190	0.05274	0.4865
	a_4	-0.00557	-0.0227	-0.00053	-0.0148	0.00689	-0.0746	-0.00190	-0.1761
check sum		12.17659	-3.5735	12.27328	-14.4305	12.17757	-21.9569	11.94330	-23.0237
Venus	a_0	15.05170	19.2341	14.79317	13.2236	14.51778	0.9135	14.19458	-12.4972
	a_1	-0.19087	-2.5683	-0.30177	-10.0767	-0.28193	-14.4773	-0.44781	-13.1182
	a_2	-0.17883	-4.0600	0.01442	-3.5485	-0.02496	-0.8748	-0.14739	2.3267
	a_3	0.12103	-0.2821	0.02366	0.8038	-0.04404	1.1126	-0.03543	1.3536
	a_4	-0.02867	0.2565	-0.02055	0.0580	-0.00086	0.0181	0.02335	-0.1308
check sum		14.77436	12.5802	14.50893	0.4602	14.16599	-13.3079	13.58730	-22.0659
Mars	a_0	21.79357	-0.4008	0.19304	-2.2202	2.58159	-1.9974	4.21792	2.0293
	a_1	2.21534	-0.8738	2.72315	-1.8484	2.09385	2.5405	1.45816	5.6384
	a_2	0.49756	-2.2755	-0.04420	1.3551	-0.42894	2.2900	-0.24775	1.0848
	a_3	-0.07263	1.0885	-0.33516	1.3877	0.05255	-0.4272	0.05748	-0.3724
	a_4	-0.07095	0.1323	0.10981	-0.5905	0.00908	-0.0201	-0.00737	0.0212
check sum		24.36289	-2.3293	2.64664	-1.9163	4.30813	2.3858	5.47844	8.4013
Jupiter	a_0	18.39932	20.1955	20.30460	20.3109	22.48673	19.9153	0.72782	19.1869
	a_1	1.92705	0.3939	2.14254	-0.1438	2.34768	-0.6481	2.39916	-0.8182
	a_2	0.10339	-0.2932	0.12204	-0.2888	0.08147	-0.2202	-0.02533	0.0561
	a_3	0.01308	0.0057	-0.00211	0.0073	-0.03758	0.0762	-0.04496	0.1499
	a_4	-0.00386	-0.0012	-0.00690	0.0093	-0.00065	0.0127	0.00853	-0.0228
check sum		20.43898	20.3007	22.56017	19.8949	24.87765	19.1359	3.06522	18.5519
Saturn	a_0	4.91947	-22.4080	6.83659	-22.5118	8.71842	-22.6224	10.47059	-22.6705
	a_1	2.10074	-0.0914	1.99111	-0.1229	1.89949	-0.0936	1.84506	-0.0007
	a_2	-0.06008	-0.0273	-0.05521	-0.0040	-0.03970	0.0346	-0.01982	0.0596
	a_3	-0.00104	0.0056	0.00463	0.0138	0.00789	0.0135	0.00909	0.0087
	a_4	0.00173	0.0017	0.00062	-0.0004	-0.00027	-0.0025	-0.00066	-0.0051
check sum		6.96082	-22.5194	8.77774	-22.6253	10.58583	-22.6704	12.30426	-22.6080

Semi-diameter (S) of the Sun and HP of Venus and Mars

		°	°	°	°
Sun S	a_0	0.2645	0.2667	0.2691	0.2709
	a_1	0.0023	0.0024	0.0019	0.0008
Venus HP	a_0	0.0031	0.0024	0.0020	0.0018
	a_1	-0.0007	-0.0004	-0.0003	-0.0002
Mars HP	a_0	0.0059	0.0062	0.0048	0.0034
	a_1	0.0003	-0.0014	-0.0014	-0.0010

$$GHA^\circ = 15((((a_4x + a_3)x + a_2)x + a_1)x + a_0 + GMT^h)$$

$$DEC^\circ = (((a_4x + a_3)x + a_2)x + a_1)x + a_0$$

$$S^\circ,\ HP^\circ = a_1x + a_0 \qquad \text{where } x = (d + GMT^h/24)/32$$

		January GHA-GMT (h)	January DEC (°)	February GHA-GMT (h)	February DEC (°)	March GHA-GMT (h)	March DEC (°)	April GHA-GMT (h)	April DEC (°)
Sun	a_0	11.95123	-23.1004	11.77647	-17.4480	11.78940	-8.0555	11.92827	4.0608
	a_1	-0.25485	2.3290	-0.08123	8.9036	0.09794	12.0831	0.15996	12.4053
	a_2	0.03331	3.9950	0.11533	2.6764	0.07160	1.0384	-0.01444	-0.6242
	a_3	0.06347	-0.3110	0.00387	-0.7067	-0.02031	-0.6802	-0.02375	-0.5749
	a_4	-0.01912	-0.0798	-0.01173	0.0445	-0.00539	0.0621	0.00012	0.0394
check sum		11.77404	-17.1672	11.80271	-6.5302	11.93324	4.4479	12.05016	15.3064
Venus	a_0	13.61077	-21.9065	12.85339	-21.5142	12.32591	-12.5476	11.97119	2.3317
	a_1	-0.74673	-5.2783	-0.73689	6.0877	-0.46352	13.6203	-0.32308	16.0547
	a_2	-0.12506	5.5043	0.12418	5.5915	0.14762	2.9092	-0.01326	-0.3415
	a_3	0.09706	0.7421	0.05891	-0.8862	-0.03966	-1.1891	-0.06812	-1.1201
	a_4	-0.00555	-0.3801	-0.02939	-0.0808	-0.00928	0.0403	0.00896	0.0031
check sum		12.83049	-21.3185	12.27020	-10.8020	11.96107	2.8331	11.57569	16.9279
Mars	a_0	5.44376	8.1898	6.41857	14.6822	7.16005	19.7610	7.86945	23.5734
	a_1	1.11333	6.7676	0.91561	6.4026	0.78728	5.0524	0.68781	2.6905
	a_2	-0.13198	0.2008	-0.08908	-0.4842	-0.06861	-0.9909	-0.04178	-1.3929
	a_3	0.02717	-0.3062	0.01572	-0.2570	0.01357	-0.1977	0.01921	-0.1023
	a_4	-0.00518	0.0299	-0.00341	0.0330	-0.00139	0.0314	-0.00295	0.0300
check sum		6.44710	14.8819	7.25741	20.3766	7.89090	23.6562	8.53174	24.7987
Jupiter	a_0	2.99486	18.5631	5.06877	18.5783	6.75720	19.1941	8.46587	20.2708
	a_1	2.25471	-0.3717	2.02576	0.4047	1.84193	0.9542	1.69665	1.2033
	a_2	-0.11083	0.3743	-0.11889	0.3945	-0.09303	0.2360	-0.06061	0.0347
	a_3	-0.01257	0.0511	0.01003	-0.0577	0.01385	-0.0826	0.01292	-0.0693
	a_4	0.00580	-0.0256	0.00036	-0.0030	-0.00112	0.0067	-0.00168	0.0097
check sum		5.13197	18.5912	6.98603	19.3168	8.51883	20.3084	10.11315	21.4492
Saturn	a_0	12.24708	-22.6117	14.02990	-22.4456	15.67978	-22.2538	17.58397	-22.0864
	a_1	1.82983	0.1218	1.85778	0.2100	1.91831	0.2135	2.01717	0.1154
	a_2	0.00297	0.0614	0.02562	0.0279	0.04358	-0.0218	0.05815	-0.0756
	a_3	0.00877	-0.0092	0.00748	-0.0209	0.00626	-0.0245	0.00230	-0.0127
	a_4	-0.00067	-0.0013	-0.00052	0.0011	-0.00087	0.0037	-0.00106	0.0042
check sum		14.08798	-22.4390	15.92026	-22.2275	17.64706	-22.0829	19.66053	-22.0551

Semi-diameter (S) of the Sun and HP of Venus and Mars

		January (°)	February (°)	March (°)	April (°)
Sun S	a_0	0.2716	0.2710	0.2695	0.2672
	a_1	-0.0006	-0.0018	-0.0023	-0.0023
Venus HP	a_0	0.0016	0.0015	0.0014	0.0014
	a_1	-0.0001	-0.0001	0.0000	0.0000
Mars HP	a_0	0.0025	0.0019	0.0016	0.0013
	a_1	-0.0006	-0.0004	-0.0003	-0.0002

$$GHA^\circ = 15((((a_4x + a_3)x + a_2)x + a_1)x + a_0 + GMT^h)$$

$$DEC^\circ = (((a_4x + a_3)x + a_2)x + a_1)x + a_0$$

$$S^\circ,\ HP^\circ = a_1x + a_0 \qquad \text{where } x = (d + GMT^h/24)/32$$

		May		June		July		August	
		GHA-GMT h	DEC °	GHA-GMT h	DEC °	GHA-GMT h	DEC °	GHA-GMT h	DEC °
Sun	a_0	12.04607	14.6989	12.04103	21.8825	11.94163	23.1895	11.89394	18.3281
	a_1	0.07056	9.8491	-0.07715	4.6042	-0.10712	-1.8802	0.02383	-7.8635
	a_2	-0.07760	-2.0640	-0.06189	-3.2637	0.03278	-3.4936	0.09046	-2.5537
	a_3	-0.00514	-0.5226	0.03439	-0.2666	0.03921	0.2115	-0.00301	0.4570
	a_4	0.00467	0.0619	-0.00131	0.1019	-0.01174	0.0526	-0.00736	0.0002
check sum		12.03856	22.0233	11.93507	23.0583	11.89476	18.0798	11.99786	8.3681
Venus	a_0	11.60742	16.1625	10.98676	24.0237	10.30811	21.4802	9.87115	9.7469
	a_1	-0.49792	12.4647	-0.75077	2.9349	-0.62490	-8.0714	-0.29069	-15.1536
	a_2	-0.15461	-3.4381	-0.06552	-6.0363	0.16930	-5.1926	0.13585	-2.0828
	a_3	-0.01868	-1.2793	0.11795	-0.2995	0.03543	1.0426	-0.05069	1.0882
	a_4	0.02702	0.1995	-0.01866	0.3328	-0.02575	0.0126	-0.00413	-0.0375
check sum		10.96323	24.1093	10.26976	20.9556	9.86219	9.2714	9.66149	-6.4388
Mars	a_0	8.49110	24.8104	9.12984	23.3080	9.79537	19.3786	10.55085	13.3053
	a_1	0.65063	-0.0914	0.67838	-2.9655	0.74428	-5.3232	0.81152	-7.0987
	a_2	-0.00588	-1.5372	0.02655	-1.4091	0.03758	-1.1071	0.02888	-0.7254
	a_3	0.01969	0.0151	0.01176	0.1099	0.00232	0.1453	-0.00555	0.1355
	a_4	-0.00449	0.0171	-0.00449	-0.0016	-0.00333	-0.0110	-0.00166	-0.0059
check sum		9.15105	23.2140	9.84204	19.0417	10.57622	13.0826	11.38404	5.6108
Jupiter	a_0	10.01256	21.3797	11.55101	22.3236	13.02419	22.8956	14.57301	23.1036
	a_1	1.61154	1.1170	1.57213	0.8048	1.57709	0.4111	1.62805	0.0345
	a_2	-0.03333	-0.1128	-0.00813	-0.1983	0.01476	-0.2159	0.04090	-0.1646
	a_3	0.01072	-0.0433	0.00817	-0.0144	0.00760	0.0105	0.00740	0.0289
	a_4	-0.00135	0.0080	-0.00037	0.0046	0.00028	0.0033	0.00130	0.0040
check sum		11.60014	22.3486	13.12281	22.9203	14.62392	23.1046	16.25066	23.0064
Saturn	a_0	19.52725	-22.0520	21.64086	-22.1728	23.75602	-22.3805	1.94760	-22.5885
	a_1	2.12872	-0.0454	2.22872	-0.1904	2.27223	-0.2343	2.23967	-0.1853
	a_2	0.05990	-0.0931	0.04108	-0.0560	0.00391	0.0053	-0.03687	0.0417
	a_3	-0.00406	0.0097	-0.01214	0.0268	-0.01575	0.0197	-0.01040	0.0022
	a_4	-0.00127	0.0021	-0.00046	-0.0027	0.00114	-0.0044	0.00159	-0.0009
check sum		21.71054	-22.1787	23.89806	-22.3951	26.01755	-22.5942	4.14159	-22.7308

Semi-diameter (S) of the Sun and HP of Venus and Mars

		May °	June °	July °	August °
Sun S	a_0	0.2649	0.2632	0.2625	0.2630
	a_1	-0.0017	-0.0007	0.0005	0.0016
Venus HP	a_0	0.0014	0.0015	0.0016	0.0018
	a_1	0.0001	0.0001	0.0002	0.0003
Mars HP	a_0	0.0012	0.0011	0.0010	0.0009
	a_1	-0.0001	-0.0001	0.0000	0.0000

$$GHA^{\circ} = 15((((a_4 x + a_3)x + a_2)x + a_1)x + a_0 + GMT^{h})$$

$$DEC^{\circ} = (((a_4 x + a_3)x + a_2)x + a_1)x + a_0$$

$$S^{\circ},\ HP^{\circ} = a_1 x + a_0 \qquad \text{where } x = (d + GMT^{h}/24)/32$$

Table 1: SUN AND PLANETS, 1989

		September		October		November		December	
		GHA-GMT h	DEC °	GHA-GMT h	DEC °	GHA-GMT h	DEC °	GHA-GMT h	DEC °
Sun	a_0	11.99271	8.7296	12.16443	-2.7049	12.27247	-14.0304	12.19143	-21.6063
	a_1	0.16363	-11.5265	0.17356	-12.4464	0.02164	-10.4079	-0.19285	-5.2510
	a_2	0.05054	-1.1491	-0.03122	0.2761	-0.10358	1.9283	-0.09068	3.5069
	a_3	-0.03100	0.4422	-0.03878	0.4786	-0.02728	0.6100	0.02852	0.4681
	a_4	-0.00074	0.0222	0.00504	0.0429	0.01577	-0.0207	0.00876	-0.1598
check sum		12.17514	-3.4816	12.27303	-14.3537	12.17902	-21.9207	11.94518	-23.0421
Venus	a_0	9.66733	-5.9315	9.43284	-19.5229	9.03800	-26.6815	8.91431	-24.4980
	a_1	-0.18552	-16.2594	-0.34395	-11.8337	-0.39804	-2.3843	0.29759	6.2877
	a_2	-0.02500	0.9113	-0.11136	3.7465	0.12868	5.4979	0.61961	3.1006
	a_3	-0.06040	1.0462	0.01542	1.1315	0.15047	-0.1453	0.31264	-1.7003
	a_4	0.01448	-0.0143	0.03272	-0.2718	0.01620	-0.3801	0.00205	-0.1733
check sum		9.41089	-20.2477	9.02567	-26.7504	8.93531	-24.0933	10.14620	-16.9833
Mars	a_0	11.35761	5.8657	12.14805	-1.9606	12.92359	-9.9499	13.58534	-16.7552
	a_1	0.84572	-8.1448	0.83082	-8.4389	0.75980	-7.9104	0.64471	-6.4405
	a_2	0.00736	-0.3333	-0.01988	0.0642	-0.04737	0.5339	-0.06577	1.0700
	a_3	-0.01074	0.1168	-0.01254	0.1209	-0.01381	0.1530	-0.00931	0.1959
	a_4	-0.00004	0.0079	0.00083	0.0179	0.00317	0.0199	0.00537	0.0024
check sum		12.19991	-2.4877	12.94728	-10.1965	13.62538	-17.1535	14.16034	-21.9274
Jupiter	a_0	16.19644	23.0125	17.89184	22.8243	19.83165	22.7590	21.92197	22.9332
	a_1	1.73282	-0.1888	1.89420	-0.1727	2.11818	0.0624	2.33216	0.2826
	a_2	0.07108	-0.0539	0.10281	0.0758	0.12254	0.1488	0.09157	0.0621
	a_3	0.00892	0.0419	0.01033	0.0415	0.00252	-0.0055	-0.02665	-0.0666
	a_4	0.00138	0.0022	-0.00120	-0.0079	-0.00682	-0.0135	-0.00510	0.0101
check sum		18.01064	22.8139	19.89798	22.7610	22.06807	22.9512	24.31395	23.2214
Saturn	a_0	4.07461	-22.7275	6.03260	-22.7805	7.94944	-22.7347	9.72369	-22.5665
	a_1	2.14477	-0.1008	2.03223	-0.0096	1.92935	0.1096	1.86145	0.2521
	a_2	-0.05919	0.0469	-0.05948	0.0542	-0.04534	0.0709	-0.02563	0.0789
	a_3	-0.00169	-0.0012	0.00411	0.0029	0.00634	0.0038	0.00684	-0.0008
	a_4	0.00088	0.0019	0.00023	0.0017	0.00015	-0.0001	0.00024	-0.0022
check sum		6.15938	-22.7807	8.00969	-22.7313	9.83994	-22.5505	11.56659	-22.2385

Semi-diameter (S) of the Sun and HP of Venus and Mars

		September °	October °	November °	December °
Sun S	a_0	0.2645	0.2667	0.2691	0.2709
	a_1	0.0023	0.0024	0.0018	0.0008
Venus HP	a_0	0.0021	0.0025	0.0033	0.0048
	a_1	0.0005	0.0009	0.0017	0.0032
Mars HP	a_0	0.0009	0.0009	0.0009	0.0010
	a_1	0.0000	0.0000	0.0000	0.0001

$$GHA^\circ = 15((((a_4x + a_3)x + a_2)x + a_1)x + a_0 + GMT^h)$$

$$DEC^\circ = (((a_4x + a_3)x + a_2)x + a_1)x + a_0$$

$$S^\circ,\ HP^\circ = a_1x + a_0 \qquad \text{where } x = (d + GMT^h/24)/32$$

		January		February		March		April	
		GHA-GMT	DEC	GHA-GMT	DEC	GHA-GMT	DEC	GHA-GMT	DEC
		h	°	h	°	h	°	h	°
Sun	a_0	11.95315	-23.1173	11.77706	-17.5142	11.78861	-8.1457	11.92686	3.9688
	a_1	-0.25698	2.2795	-0.08356	8.8709	0.09528	12.0795	0.15923	12.4160
	a_2	0.04561	3.9383	0.12474	2.6115	0.08357	0.9570	-0.00361	-0.6683
	a_3	0.04091	-0.2135	-0.01412	-0.5606	-0.03896	-0.5184	-0.04273	-0.4564
	a_4	-0.00813	-0.1214	-0.00236	-0.0295	0.00333	-0.0163	0.00960	-0.0251
check sum		11.77456	-17.2344	11.80176	-6.6219	11.93183	4.3561	12.04935	15.2350
Venus	a_0	10.07154	-17.1954	13.13512	-14.2864	14.59771	-15.5588	14.84651	-12.3370
	a_1	2.32109	6.8604	2.87152	-0.8913	0.72583	-0.2345	-0.01710	7.1188
	a_2	1.90087	-2.6398	-1.58312	-3.1476	-0.73562	2.9607	-0.12721	3.7982
	a_3	-0.84867	-2.7072	0.15132	3.3896	0.31082	1.2259	0.08016	-0.5089
	a_4	-0.22324	1.3650	0.10303	-0.6034	-0.05289	-0.5043	-0.02387	-0.1547
check sum		13.22159	-14.3170	14.67787	-15.5391	14.84585	-12.1110	14.75849	-2.0836
Mars	a_0	14.14444	-21.8101	14.58976	-23.8181	14.95311	-22.4751	15.39459	-17.6389
	a_1	0.51048	-3.8059	0.42187	-0.2197	0.42219	3.2699	0.49790	6.5466
	a_2	-0.06055	1.6340	-0.02167	1.9859	0.02560	1.9050	0.05116	1.3985
	a_3	0.00296	0.1928	0.01451	0.0713	0.00930	-0.0841	-0.00262	-0.2073
	a_4	0.00561	-0.0339	0.00184	-0.0561	0.00001	-0.0486	0.00078	-0.0294
check sum		14.60294	-23.8231	15.00631	-22.0367	15.41021	-17.4329	15.94181	-9.9305
Jupiter	a_0	0.23846	23.2136	2.53169	23.3849	4.44366	23.4571	6.34769	23.4901
	a_1	2.41608	0.2527	2.28791	0.1115	2.07776	0.0631	1.86184	-0.0133
	a_2	-0.01466	-0.0806	-0.10855	-0.0529	-0.12410	-0.0183	-0.09489	-0.0732
	a_3	-0.04140	-0.0073	-0.01455	0.0323	0.00752	-0.0037	0.01214	-0.0286
	a_4	0.00459	0.0099	0.00495	-0.0111	0.00094	-0.0086	-0.00020	-0.0004
check sum		2.60307	23.3883	4.70145	23.4647	6.40578	23.4896	8.12658	23.3746
Saturn	a_0	11.50935	-22.2509	13.28713	-21.8228	14.92006	-21.4124	16.79242	-21.0499
	a_1	1.83191	0.3958	1.84525	0.4746	1.89240	0.4460	1.97845	0.2790
	a_2	-0.00333	0.0562	0.01826	0.0112	0.03632	-0.0520	0.05165	-0.1185
	a_3	0.00636	-0.0023	0.00614	-0.0167	0.00540	-0.0216	0.00441	-0.0212
	a_4	0.00053	-0.0067	0.00040	-0.0037	0.00013	-0.0013	-0.00086	0.0040
check sum		13.34482	-21.8079	15.15718	-21.3574	16.85431	-21.0413	18.82607	-20.9066

Semi-diameter (S) of the Sun and HP of Venus and Mars

			January	February	March	April
			°	°	°	°
Sun	S	a_0	0.2716	0.2710	0.2695	0.2672
		a_1	-0.0006	-0.0018	-0.0024	-0.0023
Venus	HP	a_0	0.0084	0.0082	0.0052	0.0035
		a_1	0.0003	-0.0033	-0.0018	-0.0009
Mars	HP	a_0	0.0011	0.0011	0.0013	0.0014
		a_1	0.0001	0.0001	0.0002	0.0002

$$GHA^\circ = 15((((a_4x + a_3)x + a_2)x + a_1)x + a_0 + GMT^h)$$

$$DEC^\circ = (((a_4x + a_3)x + a_2)x + a_1)x + a_0$$

$$S^\circ, HP^\circ = a_1x + a_0 \qquad \text{where } x = (d + GMT^h/24)/32$$

		May		June		July		August	
		GHA-GMT h	DEC °	GHA-GMT h	DEC °	GHA-GMT h	DEC °	GHA-GMT h	DEC °
Sun	a_0	12.04518	14.6260	12.04110	21.8480	11.94186	23.2022	11.89320	18.3843
	a_1	0.07097	9.8750	-0.07588	4.6436	-0.10674	-1.8333	0.02413	-7.8182
	a_2	-0.06695	-2.0673	-0.05553	-3.2225	0.03659	-3.4512	0.09056	-2.5510
	a_3	-0.02578	-0.4743	0.01680	-0.3167	0.02542	0.1179	-0.00976	0.3933
	a_4	0.01525	0.0304	0.00886	0.1218	-0.00307	0.1020	-0.00192	0.0447
check sum		12.03867	21.9898	11.93535	23.0742	11.89406	18.1376	11.99621	8.4531
Venus	a_0	14.76629	-2.8637	14.59708	9.7759	14.23253	19.7545	13.61123	22.5407
	a_1	-0.12331	12.3881	-0.25550	12.7566	-0.53362	7.6005	-0.68877	-2.3108
	a_2	-0.02443	1.6116	-0.11158	-1.2435	-0.15636	-4.1147	0.01771	-5.5930
	a_3	-0.02808	-0.9144	-0.05152	-1.1021	0.02882	-1.0709	0.08381	0.2033
	a_4	-0.00147	-0.0485	0.02009	0.0264	0.01838	0.2935	-0.02215	0.2396
check sum		14.58900	10.1731	14.19857	20.2133	13.58975	22.4629	13.00183	15.0798
Mars	a_0	15.90479	-10.4659	16.51255	-1.8084	17.15494	6.4195	17.87570	13.5159
	a_1	0.58953	8.5230	0.66113	9.0836	0.71054	8.2400	0.79237	6.2497
	a_2	0.04598	0.6880	0.02831	-0.0748	0.02486	-0.7378	0.05874	-1.1821
	a_3	-0.00988	-0.2560	-0.00655	-0.2827	0.00474	-0.2557	0.03034	-0.1470
	a_4	0.00283	-0.0137	0.00402	0.0138	0.00545	0.0440	0.00495	0.0768
check sum		16.53325	-1.5246	17.19946	6.9315	17.90053	13.7100	18.76210	18.5133
Jupiter	a_0	8.01961	23.3894	9.63425	23.0075	11.14105	22.2883	12.68653	21.1975
	a_1	1.71515	-0.2272	1.62766	-0.5782	1.59407	-0.9562	1.60372	-1.2749
	a_2	-0.05967	-0.1509	-0.02891	-0.1950	-0.00506	-0.1853	0.01601	-0.1268
	a_3	0.00993	-0.0271	0.00693	-0.0155	0.00562	-0.0012	0.00630	0.0286
	a_4	0.00006	0.0050	0.00073	0.0090	0.00099	0.0120	0.00091	0.0081
check sum		9.68508	22.9892	11.24066	22.2278	12.73667	21.1576	14.31347	19.8325
Saturn	a_0	18.69558	-20.9069	20.76832	-21.0348	22.85976	-21.3707	1.05543	-21.7753
	a_1	2.08417	0.0132	2.19240	-0.2684	2.26029	-0.4196	2.25913	-0.3843
	a_2	0.05807	-0.1518	0.04852	-0.1211	0.01959	-0.0311	-0.02092	0.0638
	a_3	0.00110	-0.0051	-0.00568	0.0218	-0.01280	0.0349	-0.01374	0.0220
	a_4	-0.00203	0.0074	-0.00246	0.0054	-0.00083	-0.0007	0.00142	-0.0027
check sum		20.83689	-21.0432	23.00110	-21.3971	25.12601	-21.7872	3.28132	-22.0765

Semi-diameter (S) of the Sun and HP of Venus and Mars

		May °	June °	July °	August °
Sun S	a_0	0.2650	0.2632	0.2626	0.2630
	a_1	-0.0017	-0.0007	0.0005	0.0016
Venus HP	a_0	0.0026	0.0021	0.0018	0.0016
	a_1	-0.0005	-0.0003	-0.0002	-0.0001
Mars HP	a_0	0.0016	0.0018	0.0020	0.0024
	a_1	0.0002	0.0003	0.0004	0.0005

$$GHA^\circ = 15((((a_4x + a_3)x + a_2)x + a_1)x + a_0 + GMT^h)$$

$$DEC^\circ = (((a_4x + a_3)x + a_2)x + a_1)x + a_0$$

$$S^\circ,\ HP^\circ = a_1x + a_0 \qquad \text{where } x = (d + GMT^h/24)/32$$

		September GHA-GMT h	September DEC °	October GHA-GMT h	October DEC °	November GHA-GMT h	November DEC °	December GHA-GMT h	December DEC °
Sun	a_0	11.99100	8.8132	12.16272	-2.6135	12.27206	-13.9522	12.19282	-21.5655
	a_1	0.16503	-11.4902	0.17645	-12.4284	0.02641	-10.4197	-0.18793	-5.2996
	a_2	0.04777	-1.2002	-0.03490	0.2034	-0.11215	1.8489	-0.10255	3.4686
	a_3	-0.03227	0.4548	-0.03882	0.5298	-0.02055	0.6932	0.03853	0.5216
	a_4	0.00208	0.0331	0.00733	0.0328	0.01490	-0.0532	0.00649	-0.1856
check sum		12.17361	-3.3893	12.27278	-14.2759	12.18067	-21.8830	11.94736	-23.0605
Venus	a_0	13.01729	15.4490	12.64236	2.0698	12.27830	-12.9420	11.69315	-22.7361
	a_1	-0.49959	-11.7033	-0.33648	-15.8811	-0.46943	-14.0038	-0.77021	-5.8914
	a_2	0.13472	-3.7319	0.02150	-0.7048	-0.14392	2.6163	-0.13621	5.7591
	a_3	-0.01961	1.0301	-0.07042	1.1043	-0.05279	1.3738	0.09616	0.8484
	a_4	-0.01141	0.0308	0.00653	0.0349	0.03232	-0.1258	0.00064	-0.4267
check sum		12.62140	1.0747	12.26349	-13.3769	11.64448	-23.0815	10.88353	-22.4467
Mars	a_0	18.73035	18.3949	19.88050	21.0639	21.77479	22.5174	0.42265	22.5770
	a_1	1.01044	3.8266	1.51295	2.0408	2.45510	0.9479	2.95744	-0.8221
	a_2	0.16211	-1.1828	0.34807	-0.7066	0.56521	-0.5628	-0.14851	-0.7981
	a_3	0.07085	0.1017	0.15612	0.2971	-0.09514	-0.7794	-0.32856	1.2924
	a_4	0.00266	0.0484	-0.04552	-0.1485	-0.09326	0.3993	0.11224	-0.3103
check sum		19.97641	21.1888	21.85212	22.5467	24.60670	22.5224	3.01526	21.9389
Jupiter	a_0	14.26169	19.8767	15.85450	18.5987	17.61903	17.6014	19.49160	17.2894
	a_1	1.65585	-1.4094	1.75070	-1.2589	1.90248	-0.7308	2.09902	0.1031
	a_2	0.03739	-0.0086	0.06196	0.1606	0.09107	0.3626	0.11123	0.4830
	a_3	0.00889	0.0585	0.01205	0.0834	0.01415	0.0888	0.00602	0.0204
	a_4	0.00034	0.0034	-0.00064	-0.0049	-0.00351	-0.0241	-0.00755	-0.0409
check sum		15.96416	18.5206	17.67857	17.5789	19.62322	17.2979	21.70032	17.8550
Saturn	a_0	3.21310	-22.0701	5.21213	-22.1656	7.16896	-22.0442	8.97205	-21.7137
	a_1	2.18508	-0.2090	2.07653	0.0093	1.96620	0.2427	1.88605	0.4597
	a_2	-0.05118	0.1085	-0.05963	0.1168	-0.05075	0.1163	-0.03324	0.1086
	a_3	-0.00758	0.0081	-0.00049	0.0058	0.00425	0.0046	0.00657	-0.0039
	a_4	0.00226	-0.0021	0.00181	-0.0027	0.00115	-0.0040	0.00056	-0.0039
check sum		5.34168	-22.1646	7.23035	-22.0364	9.08981	-21.6846	10.83199	-21.1532

Semi-diameter (S) of the Sun and HP of Venus and Mars

		September °	October °	November °	December °
Sun S	a_0	0.2645	0.2667	0.2690	0.2708
	a_1	0.0023	0.0024	0.0019	0.0008
Venus HP	a_0	0.0015	0.0014	0.0014	0.0014
	a_1	-0.0001	0.0000	0.0000	0.0001
Mars HP	a_0	0.0029	0.0036	0.0045	0.0047
	a_1	0.0007	0.0009	0.0001	-0.0011

$$GHA^\circ = 15((((a_4x + a_3)x + a_2)x + a_1)x + a_0 + GMT^h)$$

$$DEC^\circ = (((a_4x + a_3)x + a_2)x + a_1)x + a_0$$

$$S^\circ,\ HP^\circ = a_1x + a_0 \qquad \text{where } x = (d + GMT^h/24)/32$$

Table 2: MOON, 1986

Day		January GHA °	January DEC °	January HP °	February GHA °	February DEC °	February HP °	March GHA °	March DEC °	March HP °
1	a_0	301.7609	13.8046	0.9504	286.2440	-9.5648	0.9806	303.9878	-14.3667	0.9899
	a_1	348.6672	-5.4089	0.0093	348.3096	-6.1947	0.0045	347.5865	-5.8866	-0.0002
	a_2	0.1280	-0.2949		-0.3547	0.2896		-0.3893	0.4722	
2	a_0	290.5535	8.1033	0.9597	274.1968	-15.4660	0.9850	291.1846	-19.7772	0.9897
	a_1	348.9274	-5.9996	0.0097	347.5913	-5.6175	0.0034	346.7934	-4.9396	-0.0018
	a_2	0.0001	-0.1707		-0.4564	0.4771		-0.4088	0.6567	
3	a_0	279.4778	1.9360	0.9694	261.3315	-20.6025	0.9885	277.5713	-24.0569	0.9879
	a_1	348.9297	-6.3431	0.0098	346.6616	-4.6624	0.0021	345.9571	-3.6188	-0.0031
	a_2	-0.1508	-0.0351		-0.4700	0.6713		-0.3119	0.8100	
4	a_0	268.2534	-4.4389	0.9793	247.5257	-24.5902	0.9906	263.2209	-26.8639	0.9848
	a_1	348.6279	-6.4171	0.0094	345.6997	-3.3122	0.0003	345.3192	-1.9860	-0.0042
	a_2	-0.3104	0.1226		-0.3465	0.8390		-0.0963	0.8952	
5	a_0	256.5679	-10.7294	0.9887	232.8842	-27.0617	0.9909	248.4493	-27.9549	0.9807
	a_1	348.0022	-6.1766	0.0082	344.9914	-1.6193	-0.0018	345.1269	-0.1807	-0.0051
	a_2	-0.4554	0.3122		-0.0868	0.9293		0.1716	0.8839	
6	a_0	244.1130	-16.5892	0.9969	217.7951	-27.7521	0.9891	233.7522	-27.2538	0.9756
	a_1	347.0783	-5.5553	0.0060	344.8205	0.2570	-0.0042	345.4845	1.5987	-0.0061
	a_2	-0.5410	0.5328		0.2211	0.9040		0.3885	0.7791	
7	a_0	230.6512	-21.6072	1.0030	202.8415	-26.5938	0.9849	219.6273	-24.8794	0.9695
	a_1	345.9730	-4.4869	0.0030	345.2815	2.0779	-0.0067	346.2799	3.1622	-0.0070
	a_2	-0.5014	0.7587		0.4528	0.7699		0.4893	0.6148	
8	a_0	216.1272	-25.3318	1.0060	188.5777	-23.7500	0.9782	206.3964	-21.1062	0.9625
	a_1	344.9447	-2.9565	-0.0008	346.2083	3.6223	-0.0089	347.2724	4.3917	-0.0079
	a_2	-0.2873	0.9324		0.5431	0.5743		0.4776	0.4320	
9	a_0	200.7918	-27.3548	1.0052	175.3284	-19.5577	0.9693	194.1447	-16.2861	0.9546
	a_1	344.3599	-1.0705	-0.0048	347.3087	4.7688	-0.0105	348.2348	5.2527	-0.0085
	a_2	0.0597	0.9864		0.5099	0.3697		0.3935	0.2589	
10	a_0	185.2184	-27.4407	1.0004	163.1449	-14.4229	0.9587	182.7706	-10.7776	0.9461
	a_1	344.4935	0.9220	-0.0086	348.3350	5.5033	-0.0114	349.0243	5.7668	-0.0089
	a_2	0.3956	0.8952		0.4054	0.1887		0.2766	0.1077	
11	a_0	170.1115	-25.6274	0.9917	151.8826	-8.7339	0.9472	172.0688	-4.9058	0.9371
	a_1	345.3109	2.7219	-0.0116	349.1475	5.8756	-0.0115	349.5773	5.9789	-0.0089
	a_2	0.5868	0.7004		0.2732	0.0418		0.1515	-0.0202	
12	a_0	156.0098	-22.2097	0.9800	141.3004	-2.8190	0.9357	161.7950	1.0506	0.9282
	a_1	346.5061	4.1222	-0.0136	349.6932	5.9551	-0.0106	349.8790	5.9359	-0.0083
	a_2	0.6096	0.4747		0.1371	-0.0741		0.0302	-0.1291	
13	a_0	143.1237	-17.6171	0.9664	131.1280	3.0600	0.9251	151.7019	6.8554	0.9199
	a_1	347.7365	5.0657	-0.0142	349.9659	5.8044	-0.0090	349.9375	5.6762	-0.0072
	a_2	0.5198	0.2714		0.0076	-0.1680		-0.0811	-0.2249	
14	a_0	131.3771	-12.2833	0.9521	121.0990	8.6947	0.9161	141.5562	12.3049	0.9127
	a_1	348.7796	5.6019	-0.0137	349.9789	5.4673	-0.0068	349.7724	5.2257	-0.0055
	a_2	0.3805	0.1109		-0.1106	-0.2505		-0.1771	-0.3139	
15	a_0	120.5341	-6.5731	0.9384	110.9652	13.9098	0.9093	131.1500	17.2148	0.9072
	a_1	349.5403	5.8184	-0.0122	349.7545	4.9665	-0.0042	349.4137	4.5976	-0.0033
	a_2	0.2301	-0.0097		-0.2118	-0.3310		-0.2491	-0.4009	

$GHA^\circ, DEC^\circ = (a_2 x + a_1)x + a_0$ $\quad HP^\circ = a_1 x + a_0$ $\quad$ where $x = GMT^h/24$

Day		January GHA °	January DEC °	January HP °	February GHA °	February DEC °	February HP °	March GHA °	March DEC °	March HP °
16	a_0	110.3015	-0.7663	0.9261	100.5063	18.5436	0.9051	120.3139	21.4097	0.9039
	a_1	349.9989	5.7958	-0.0100	349.3258	4.3053	-0.0013	348.9091	3.7955	-0.0008
	a_2	0.0851	-0.1022		-0.2856	-0.4157		-0.2844	-0.4874	
17	a_0	100.3827	4.9257	0.9161	89.5459	22.4313	0.9038	108.9389	24.7162	0.9031
	a_1	350.1672	5.5902	-0.0073	348.7469	3.4743	0.0017	348.3322	2.8195	0.0020
	a_2	-0.0487	-0.1804		-0.3162	-0.5056		-0.2694	-0.5696	
18	a_0	90.4987	10.3340	0.9088	77.9772	25.3981	0.9055	97.0031	26.9647	0.9051
	a_1	350.0672	5.2297	-0.0044	348.1049	2.4618	0.0045	347.7856	1.6774	0.0049
	a_2	-0.1682	-0.2564		-0.2871	-0.5940		-0.1984	-0.6383	
19	a_0	80.3957	15.3056	0.9044	65.7969	27.2644	0.9100	84.5926	28.0029	0.9101
	a_1	349.7268	4.7182	-0.0016	347.5220	1.2699	0.0070	347.3845	0.3960	0.0077
	a_2	-0.2663	-0.3390		-0.1932	-0.6661		-0.0842	-0.6806	
20	a_0	69.8549	19.6828	0.9028	53.1286	27.8674	0.9171	71.8954	27.7182	0.9178
	a_1	349.1876	4.0415	0.0012	347.1319	-0.0687	0.0091	347.2177	-0.9712	0.0103
	a_2	-0.3292	-0.4321		-0.0540	-0.7036		0.0402	-0.6861	
21	a_0	58.7132	23.2902	0.9040	40.2094	27.0954	0.9262	59.1553	26.0617	0.9281
	a_1	348.5198	3.1775	0.0036	347.0273	-1.4834	0.0106	347.3050	-2.3492	0.0123
	a_2	-0.3372	-0.5317		0.0873	-0.6930		0.1351	-0.6508	
22	a_0	46.8971	25.9342	0.9077	27.3260	24.9202	0.9368	46.5962	23.0632	0.9404
	a_1	347.8346	2.1116	0.0056	347.2109	-2.8762	0.0112	347.5845	-3.6559	0.0136
	a_2	-0.2738	-0.6243		0.1851	-0.6319		0.1738	-0.5777	
23	a_0	34.4606	27.4201	0.9133	14.7227	21.4142	0.9481	34.3540	18.8318	0.9540
	a_1	347.2789	0.8571	0.0072	347.5917	-4.1452	0.0111	347.9406	-4.8155	0.0138
	a_2	-0.1427	-0.6889		0.2149	-0.5288		0.1506	-0.4721	
24	a_0	21.6002	27.5880	0.9206	2.5285	16.7429	0.9592	22.4437	13.5469	0.9679
	a_1	346.9927	-0.5288	0.0082	348.0301	-5.2062	0.0101	348.2482	-5.7636	0.0129
	a_2	0.0214	-0.7054		0.1777	-0.3958		0.0756	-0.3378	
25	a_0	8.6172	26.3546	0.9288	350.7345	11.1440	0.9694	10.7652	7.4488	0.9809
	a_1	347.0426	-1.9475	0.0087	348.3914	-6.0000	0.0084	348.4033	-6.4429	0.0109
	a_2	0.1641	-0.6654		0.0898	-0.2434		-0.0356	-0.1763	
26	a_0	355.8256	23.7435	0.9375	339.2131	4.9040	0.9778	359.1302	0.8335	0.9918
	a_1	347.3822	-3.2842	0.0087	348.5742	-6.4882	0.0063	348.3334	-6.7990	0.0078
	a_2	0.2422	-0.5770		-0.0302	-0.0779		-0.1661	0.0111	
27	a_0	343.4500	19.8848	0.9462	327.7543	-1.6584	0.9841	347.2947	-5.9500	0.9996
	a_1	347.8774	-4.4413	0.0083	348.5145	-6.6451	0.0040	347.9989	-6.7792	0.0041
	a_2	0.2429	-0.4570		-0.1647	0.0977		-0.2954	0.2206	
28	a_0	331.5690	14.9893	0.9545	316.1014	-8.2020	0.9881	334.9962	-12.5039	1.0037
	a_1	348.3711	-5.3564	0.0077	348.1824	-6.4507	0.0018	347.4003	-6.3383	0.0003
	a_2	0.1785	-0.3217		-0.2940	0.2821		-0.3926	0.4423	
29	a_0	320.1163	9.3142	0.9622				322.0035	-18.3955	1.0040
	a_1	348.7330	-6.0001	0.0069				346.6002	-5.4501	-0.0032
	a_2	0.0692	-0.1804					-0.4129	0.6564	
30	a_0	308.9157	3.1368	0.9691				308.1929	-23.1857	1.0008
	a_1	348.8741	-6.3615	0.0061				345.7548	-4.1283	-0.0059
	a_2	-0.0667	-0.0346					-0.3122	0.8296	
31	a_0	297.7200	-3.2561	0.9752				293.6400	-26.4824	0.9948
	a_1	348.7411	-6.4322	0.0053				345.1157	-2.4545	-0.0078
	a_2	-0.2142	0.1199					-0.0871	0.9219	

Table 2: MOON, 1986

Day		April GHA °	April DEC °	April HP °	May GHA °	May DEC °	May HP °	June GHA °	June DEC °	June HP °
1	a_0	278.6743	-28.0152	0.9870	266.1633	-23.1655	0.9688	250.3471	-3.4896	0.9312
	a_1	344.9421	-0.5948	-0.0089	346.8038	3.8861	-0.0127	349.8775	5.8965	-0.0104
	a_2	0.1916	0.9077		0.5780	0.4853		0.1261	-0.0514	
2	a_0	263.8126	-27.7045	0.9781	253.5432	-18.7979	0.9560	240.3478	2.3537	0.9208
	a_1	345.3407	1.2320	-0.0093	347.9686	4.8518	-0.0118	350.1271	5.7925	-0.0082
	a_2	0.4144	0.7977		0.4787	0.3082		-0.0103	-0.1394	
3	a_0	249.5698	-25.6782	0.9687	241.9877	-13.6409	0.9442	230.4621	8.0051	0.9126
	a_1	346.1887	2.8318	-0.0093	348.9284	5.4635	-0.0105	350.1034	5.5138	-0.0062
	a_2	0.5143	0.6326		0.3421	0.1635		-0.1286	-0.2245	
4	a_0	236.2726	-22.2175	0.9594	231.2552	-8.0163	0.9338	220.4348	13.2925	0.9064
	a_1	347.2315	4.0960	-0.0090	349.6119	5.7873	-0.0091	349.8419	5.0656	-0.0043
	a_2	0.4978	0.4553		0.1998	0.0454		-0.2230	-0.3134	
5	a_0	224.0001	-17.6697	0.9503	221.0641	-2.1856	0.9247	210.0525	18.0427	0.9022
	a_1	348.2342	5.0034	-0.0087	350.0096	5.8764	-0.0076	349.3895	4.4393	-0.0025
	a_2	0.4085	0.2915		0.0655	-0.0561		-0.2824	-0.4078	
6	a_0	212.6402	-12.3777	0.9417	211.1367	3.6326	0.9171	199.1594	22.0722	0.8997
	a_1	349.0533	5.5832	-0.0082	350.1383	5.7635	-0.0063	348.8162	3.6229	-0.0009
	a_2	0.2876	0.1492		-0.0555	-0.1504		-0.2912	-0.5027	
7	a_0	201.9785	-6.6479	0.9335	201.2172	9.2438	0.9108	187.6856	25.1907	0.8988
	a_1	349.6282	5.8789	-0.0077	350.0242	5.4628	-0.0049	348.2245	2.6146	0.0006
	a_2	0.1602	0.0262		-0.1592	-0.2439		-0.2370	-0.5864	
8	a_0	191.7643	-0.7451	0.9258	191.0806	14.4607	0.9059	175.6754	27.2177	0.8994
	a_1	349.9472	5.9296	-0.0071	349.7014	4.9753	-0.0036	347.7434	1.4365	0.0021
	a_2	0.0382	-0.0828		-0.2382	-0.3398		-0.1242	-0.6428	
9	a_0	181.7473	5.0997	0.9187	180.5429	19.0942	0.9023	163.2976	28.0111	0.9016
	a_1	350.0215	5.7630	-0.0063	349.2186	4.2953	-0.0022	347.4940	0.1445	0.0037
	a_2	-0.0729	-0.1834		-0.2805	-0.4369		0.0184	-0.6586	
10	a_0	171.6939	10.6773	0.9125	169.4811	22.9507	0.9000	150.8126	27.4976	0.9052
	a_1	349.8728	5.3958	-0.0052	348.6493	3.4200	-0.0007	347.5365	-1.1786	0.0053
	a_2	-0.1677	-0.2798		-0.2722	-0.5285		0.1456	-0.6310	
11	a_0	161.3975	15.7913	0.9073	157.8597	25.8406	0.8993	138.4963	25.6892	0.9105
	a_1	349.5329	4.8358	-0.0038	348.0967	2.3594	0.0010	347.8374	-2.4446	0.0071
	a_2	-0.2385	-0.3743		-0.2057	-0.6035		0.2197	-0.5689	
12	a_0	150.6912	20.2509	0.9034	145.7530	27.5956	0.9003	126.5535	22.6774	0.9176
	a_1	349.0497	4.0864	-0.0021	347.6802	1.1472	0.0029	348.2865	-3.5842	0.0089
	a_2	-0.2732	-0.4663		-0.0915	-0.6488		0.2262	-0.4865	
13	a_0	139.4680	23.8693	0.9013	133.3444	28.0938	0.9032	115.0651	18.6087	0.9265
	a_1	348.4955	3.1521	0.0000	347.4981	-0.1562	0.0051	348.7466	-4.5578	0.0108
	a_2	-0.2595	-0.5506		0.0388	-0.6562		0.1718	-0.3955	
14	a_0	127.7053	26.4694	0.9013	120.8835	27.2820	0.9082	103.9814	13.6574	0.9373
	a_1	347.9690	2.0476	0.0024	347.5821	-1.4735	0.0074	349.0957	-5.3494	0.0126
	a_2	-0.1923	-0.6185		0.1441	-0.6265		0.0712	-0.2997	
15	a_0	115.4843	27.8976	0.9037	108.6107	25.1832	0.9156	93.1455	8.0106	0.9499
	a_1	347.5804	0.8059	0.0050	347.8795	-2.7297	0.0098	349.2420	-5.9506	0.0140
	a_2	-0.0836	-0.6595		0.1945	-0.5685		-0.0623	-0.1947	

GHA°, DEC° = $(a_2x + a_1)x + a_0$ HP° = $a_1x + a_0$ where $x = GMT^h/24$

		April			May			June		
Day		GHA °	DEC °	HP °	GHA °	DEC °	HP °	GHA °	DEC °	HP °
16	a_0	102.9837	28.0438	0.9088	96.6845	21.8866	0.9255	82.3219	1.8678	0.9640
	a_1	347.4147	-0.5183	0.0078	348.2775	-3.8685	0.0122	349.1199	-6.3440	0.0147
	a_2	0.0358	-0.6662		0.1819	-0.4929		-0.2190	-0.0694	
17	a_0	90.4361	26.8598	0.9165	85.1425	17.5270	0.9377	71.2193	-4.5422	0.9787
	a_1	347.4927	-1.8555	0.0104	348.6482	-4.8558	0.0142	348.6821	-6.4888	0.0144
	a_2	0.1283	-0.6385		0.1139	-0.4056		-0.3876	0.0913	
18	a_0	78.0579	24.3670	0.9270	73.9023	12.2678	0.9519	59.5106	-10.9353	0.9932
	a_1	347.7580	-3.1362	0.0128	348.8811	-5.6693	0.0156	347.9018	-6.3134	0.0128
	a_2	0.1678	-0.5814		0.0040	-0.3046		-0.5430	0.2998	
19	a_0	65.9833	20.6511	0.9398	62.7845	6.2965	0.9676	46.8677	-16.9436	1.0061
	a_1	348.1019	-4.3021	0.0146	348.8925	-6.2826	0.0160	346.8003	-5.7187	0.0098
	a_2	0.1475	-0.5008		-0.1360	-0.1809		-0.6316	0.5541	
20	a_0	54.2312	15.8503	0.9545	51.5377	-0.1636	0.9837	33.0378	-22.1028	1.0159
	a_1	348.4036	-5.3070	0.0156	348.6219	-6.6504	0.0151	345.5091	-4.6070	0.0054
	a_2	0.0753	-0.3983		-0.2943	-0.0213		-0.5700	0.8189	
21	a_0	42.7078	10.1478	0.9702	39.8621	-6.8310	0.9989	17.9824	-25.8869	1.0214
	a_1	348.5587	-6.1077	0.0153	348.0308	-6.7001	0.0126	344.3390	-2.9523	0.0002
	a_2	-0.0360	-0.2697		-0.4514	0.1858		-0.2991	1.0168	
22	a_0	31.2277	3.7738	0.9856	27.4391	-13.3400	1.0116	2.0309	-27.8214	1.0215
	a_1	348.4893	-6.6522	0.0136	347.1173	-6.3340	0.0086	343.7324	-0.8923	-0.0052
	a_2	-0.1725	-0.1075		-0.5665	0.4401		0.1168	1.0628	
23	a_0	19.5415	-2.9817	0.9993	13.9900	-19.2283	1.0202	345.8878	-27.6534	1.0163
	a_1	348.1437	-6.8729	0.0105	345.9619	-5.4528	0.0035	343.9874	1.2555	-0.0100
	a_2	-0.3172	0.0940		-0.5693	0.7146		0.4885	0.9359	
24	a_0	7.3654	-9.7557	1.0098	359.3863	-23.9619	1.0237	330.3676	-25.4668	1.0062
	a_1	347.5032	-6.6892	0.0062	344.7937	-4.0110	-0.0021	344.9964	3.1356	-0.0135
	a_2	-0.4414	0.3325		-0.3876	0.9456		0.6683	0.7013	
25	a_0	354.4261	-16.1071	1.0159	343.7998	-27.0250	1.0216	316.0322	-21.6351	0.9926
	a_1	346.6056	-6.0241	0.0012	344.0007	-2.0962	-0.0073	346.3557	4.5351	-0.0155
	a_2	-0.4956	0.5889		-0.0306	1.0514		0.6563	0.4493	
26	a_0	340.5378	-21.5375	1.0171	327.7778	-28.0711	1.0142	303.0416	-16.6551	0.9771
	a_1	345.5912	-4.8382	-0.0037	343.9494	0.0301	-0.0115	347.6781	5.4260	-0.0160
	a_2	-0.4176	0.8211		0.3588	0.9899		0.5333	0.2351	
27	a_0	325.7163	-25.5516	1.0134	312.0912	-27.0549	1.0026	291.2496	-10.9974	0.9610
	a_1	344.7345	-3.1791	-0.0078	344.6956	2.0225	-0.0142	348.7464	5.8888	-0.0153
	a_2	-0.1786	0.9691		0.6090	0.8001		0.3713	0.0724	
28	a_0	310.2790	-27.7612	1.0055	297.3971	-24.2371	0.9884	280.3640	-5.0387	0.9457
	a_1	344.3734	-1.2202	-0.0108	345.9397	3.6231	-0.0153	349.4876	6.0284	-0.0136
	a_2	0.1556	0.9852		0.6649	0.5655		0.2084	-0.0478	
29	a_0	294.8140	-27.9985	0.9945	284.0001	-20.0529	0.9730	270.0568	0.9399	0.9321
	a_1	344.7017	0.7657	-0.0126	347.2836	4.7483	-0.0152	349.9018	5.9299	-0.0114
	a_2	0.4444	0.8727		0.5820	0.3509		0.0586	-0.1414	
30	a_0	279.9631	-26.3639	0.9819	271.8627	-14.9573	0.9577	260.0143	6.7267	0.9207
	a_1	345.6148	2.5169	-0.0131	348.4521	5.4432	-0.0141	350.0159	5.6461	-0.0089
	a_2	0.5855	0.6855		0.4365	0.1803		-0.0734	-0.2231	
31	a_0				260.7481	-9.3365	0.9436			
	a_1				349.3248	5.7985	-0.0124			
	a_2				0.2774	0.0506				

Table 2: MOON, 1986

Day		July GHA °	July DEC °	July HP °	August GHA °	August DEC °	August HP °	September GHA °	September DEC °	September HP °
1	a_0	249.9546	12.1480	0.9118	235.9489	26.4915	0.9019	217.1205	25.2558	0.9182
	a_1	349.8654	5.2002	-0.0064	347.9445	2.0119	0.0014	347.7456	-2.8117	0.0085
	a_2	-0.1831	-0.3040		-0.2002	-0.6219		0.2058	-0.5812	
2	a_0	239.6353	17.0424	0.9055	223.6958	27.8805	0.9033	205.0722	21.8648	0.9267
	a_1	349.4935	4.5929	-0.0039	347.5383	0.7627	0.0035	348.1660	-3.9775	0.0095
	a_2	-0.2615	-0.3902		-0.0779	-0.6643		0.2159	-0.4904	
3	a_0	228.8667	21.2433	0.9016	211.1592	27.9789	0.9068	193.4531	17.3992	0.9362
	a_1	348.9625	3.8128	-0.0017	347.3832	-0.5726	0.0052	348.6049	-4.9605	0.0098
	a_2	-0.2937	-0.4808		0.0642	-0.6653		0.1703	-0.3801	
4	a_0	217.5361	24.5734	0.8999	198.6090	26.7419	0.9120	182.2265	12.0612	0.9461
	a_1	348.3654	2.8494	0.0003	347.5185	-1.9092	0.0065	348.9506	-5.7223	0.0095
	a_2	-0.2640	-0.5674		0.1807	-0.6217		0.0826	-0.2556	
5	a_0	205.6394	26.8541	0.9002	186.3095	24.2126	0.9185	171.2572	6.0861	0.9555
	a_1	347.8285	1.7104	0.0020	347.8898	-3.1569	0.0074	349.1189	-6.2352	0.0086
	a_2	-0.1686	-0.6342		0.2392	-0.5417		-0.0338	-0.1187	
6	a_0	193.3023	27.9296	0.9022	174.4383	20.5162	0.9259	160.3396	-0.2647	0.9642
	a_1	347.4876	0.4355	0.0035	348.3772	-4.2425	0.0080	349.0528	-6.4746	0.0075
	a_2	-0.0276	-0.6644		0.2317	-0.4386		-0.1670	0.0317	
7	a_0	180.7653	27.7010	0.9058	163.0460	15.8374	0.9339	149.2225	-6.7041	0.9717
	a_1	347.4359	-0.9001	0.0048	348.8475	-5.1207	0.0083	348.7179	-6.4135	0.0061
	a_2	0.1157	-0.6485		0.1687	-0.3241		-0.3040	0.1983	
8	a_0	168.3190	26.1536	0.9106	152.0600	10.3951	0.9421	137.6340	-12.9156	0.9778
	a_1	347.6762	-2.2023	0.0060	349.1895	-5.7693	0.0084	348.1048	-6.0191	0.0047
	a_2	0.2162	-0.5902		0.0660	-0.2041		-0.4224	0.3822	
9	a_0	156.2122	23.3628	0.9166	141.3128	4.4244	0.9506	125.3152	-18.5484	0.9825
	a_1	348.1190	-3.3856	0.0071	349.3245	-6.1784	0.0084	347.2481	-5.2552	0.0033
	a_2	0.2498	-0.5032		-0.0631	-0.0783		-0.4837	0.5776	
10	a_0	144.5803	19.4761	0.9237	130.5712	-1.8295	0.9590	112.0807	-23.2222	0.9858
	a_1	348.6273	-4.3931	0.0082	349.1997	-6.3369	0.0084	346.2616	-4.0959	0.0020
	a_2	0.2176	-0.4025		-0.2085	0.0584		-0.4370	0.7638	
11	a_0	133.4235	14.6827	0.9319	119.5593	-8.1048	0.9674	97.9091	-26.5515	0.9878
	a_1	349.0686	-5.1982	0.0094	348.7818	-6.2234	0.0081	345.3680	-2.5575	0.0007
	a_2	0.1339	-0.2974		-0.3575	0.2140		-0.2512	0.9008	
12	a_0	122.6235	9.1894	0.9413	107.9810	-14.1104	0.9756	83.0318	-28.2073	0.9885
	a_1	349.3405	-5.7935	0.0104	348.0610	-5.7991	0.0077	344.8583	-0.7397	-0.0007
	a_2	0.0142	-0.1895		-0.4857	0.3953		0.0350	0.9443	
13	a_0	111.9752	3.2088	0.9517	95.5551	-19.5100	0.9832	67.9307	-28.0041	0.9878
	a_1	349.3718	-6.1743	0.0113	347.0755	-5.0105	0.0068	344.9391	1.1645	-0.0023
	a_2	-0.1303	-0.0731		-0.5473	0.5992		0.3111	0.8757	
14	a_0	101.2134	-3.0358	0.9630	82.0847	-23.9172	0.9901	53.1844	-25.9671	0.9855
	a_1	349.1125	-6.3241	0.0118	345.9580	-3.8082	0.0054	345.5820	2.9253	-0.0040
	a_2	-0.2905	0.0628		-0.4808	0.8008		0.4753	0.7189	
15	a_0	90.0320	-9.2935	0.9749	67.5666	-26.9216	0.9955	39.2423	-22.3270	0.9815
	a_1	348.5295	-6.2038	0.0117	344.9741	-2.1938	0.0033	346.5507	4.3656	-0.0058
	a_2	-0.4514	0.2311		-0.2515	0.9474		0.5031	0.5206	

$GHA°, DEC° = (a_2 x + a_1)x + a_0$ $HP° = a_1 x + a_0$ where $x = GMT^h/24$

Day		July GHA °	July DEC °	July HP °	August GHA °	August DEC °	August HP °	September GHA °	September DEC °	September HP °
16	a_0	78.1075	-15.2620	0.9866	52.2961	-28.1674	0.9988	26.2947	-17.4449	0.9757
	a_1	347.6179	-5.7472	0.0107	344.4652	-0.2794	0.0006	347.5672	5.4048	-0.0076
	a_2	-0.5785	0.4396		0.0834	0.9807		0.4324	0.3204	
17	a_0	65.1463	-20.5646	0.9974	36.8509	-27.4681	0.9993	14.2918	-11.7235	0.9681
	a_1	346.4407	-4.8698	0.0086	344.6479	1.7000	-0.0026	348.4359	6.0418	-0.0090
	a_2	-0.6082	0.6779		0.3848	0.8809		0.3131	0.1396	
18	a_0	50.9818	-24.7520	1.0061	21.8870	-24.8912	0.9968	3.0381	-5.5454	0.9591
	a_1	345.1952	-3.5062	0.0055	345.4422	3.4711	-0.0059	349.0624	6.3170	-0.0101
	a_2	-0.4652	0.9003		0.5419	0.6849		0.1799	-0.0158	
19	a_0	35.7185	-27.3553	1.0116	7.8712	-20.7399	0.9909	352.2776	0.7531	0.9490
	a_1	344.2432	-1.6860	0.0014	346.5450	4.8414	-0.0089	349.4205	6.2821	-0.0105
	a_2	-0.1366	1.0292		0.5456	0.4548		0.0510	-0.1475	
20	a_0	19.8332	-28.0126	1.0130	354.9599	-15.4482	0.9820	341.7467	6.8853	0.9385
	a_1	343.9744	0.3968	-0.0032	347.6458	5.7464	-0.0112	349.5198	5.9847	-0.0101
	a_2	0.2589	1.0024		0.4509	0.2382		-0.0639	-0.2605	
21	a_0	4.0723	-26.6169	1.0098	343.0538	-9.4674	0.9707	331.2007	12.6075	0.9283
	a_1	344.5185	2.4182	-0.0076	348.5504	6.2170	-0.0126	349.3881	5.4621	-0.0091
	a_2	0.5438	0.8297		0.3145	0.0568		-0.1571	-0.3602	
22	a_0	349.1367	-23.3740	1.0022	331.9157	-3.1965	0.9581	320.4305	17.7074	0.9192
	a_1	345.6337	4.0817	-0.0113	349.1787	6.3254	-0.0129	349.0685	4.7403	-0.0074
	a_2	0.6363	0.5841		0.1707	-0.0874		-0.2185	-0.4507	
23	a_0	335.4055	-18.7133	0.9908	321.2621	3.0391	0.9451	309.2800	21.9952	0.9117
	a_1	346.9227	5.2454	-0.0138	349.5179	6.1469	-0.0123	348.6241	3.8372	-0.0052
	a_2	0.5759	0.3405		0.0356	-0.2018		-0.2361	-0.5323	
24	a_0	322.9013	-13.1316	0.9770	310.8131	8.9822	0.9328	297.6688	25.2987	0.9065
	a_1	348.0805	5.9190	-0.0149	349.5861	5.7412	-0.0108	348.1439	2.7701	-0.0026
	a_2	0.4394	0.1382		-0.0830	-0.2968		-0.2004	-0.6005	
25	a_0	311.4180	-7.0775	0.9620	300.3143	14.4248	0.9220	285.6141	27.4674	0.9039
	a_1	348.9597	6.1887	-0.0148	349.4159	5.1465	-0.0086	347.7369	1.5652	0.0002
	a_2	0.2815	-0.0160		-0.1778	-0.3821		-0.1141	-0.6472	
26	a_0	300.6560	-0.9073	0.9472	289.5511	19.1875	0.9133	273.2393	28.3851	0.9042
	a_1	349.5209	6.1520	-0.0136	349.0542	4.3819	-0.0060	347.5070	0.2661	0.0031
	a_2	0.1284	-0.1319		-0.2385	-0.4635		0.0015	-0.6639	
27	a_0	290.3024	5.1109	0.9335	278.3665	23.1043	0.9073	260.7500	27.9875	0.9073
	a_1	349.7750	5.8856	-0.0116	348.5693	3.4539	-0.0032	347.5136	-1.0667	0.0060
	a_2	-0.0100	-0.2238		-0.2514	-0.5416		0.1111	-0.6470	
28	a_0	280.0649	10.7710	0.9219	266.6854	26.0152	0.9041	248.3762	26.2748	0.9133
	a_1	349.7516	5.4372	-0.0091	348.0580	2.3685	-0.0003	347.7433	-2.3648	0.0085
	a_2	-0.1286	-0.3052		-0.2063	-0.6098		0.1820	-0.5991	
29	a_0	269.6860	15.9013	0.9128	254.5390	27.7729	0.9038	236.3019	23.3125	0.9218
	a_1	349.4896	4.8270	-0.0064	347.6388	1.1447	0.0024	348.1158	-3.5660	0.0105
	a_2	-0.2200	-0.3856		-0.1076	-0.6565		0.1978	-0.5265	
30	a_0	258.9546	20.3410	0.9064	242.0729	28.2609	0.9062	224.6148	19.2219	0.9323
	a_1	349.0426	4.0561	-0.0036	347.4225	-0.1739	0.0049	348.5186	-4.6214	0.0119
	a_2	-0.2717	-0.4688		0.0196	-0.6698		0.1598	-0.4347	
31	a_0	247.7255	23.9266	0.9028	229.5173	27.4177	0.9112			
	a_1	348.4901	3.1177	-0.0010	347.4663	-1.5193	0.0070			
	a_2	-0.2682	-0.5514		0.1355	-0.6438				

Table 2: MOON, 1986

Day		October GHA °	October DEC °	October HP °	November GHA °	November DEC °	November HP °	December GHA °	December DEC °	December HP °
1	a_0	213.2914	14.1681	0.9442	200.1918	-8.4066	0.9899	193.8994	-22.9916	1.0163
	a_1	348.8433	-5.4934	0.0123	348.1604	-6.5864	0.0107	345.3056	-4.4447	0.0051
	a_2	0.0787	-0.3253		-0.4504	0.2342		-0.5548	0.8637	
2	a_0	202.2112	8.3521	0.9566	187.8998	-14.7537	1.0006	178.6566	-26.5688	1.0214
	a_1	349.0042	-6.1469	0.0119	347.2486	-6.1216	0.0070	344.1676	-2.6977	-0.0003
	a_2	-0.0328	-0.1960		-0.5462	0.4772		-0.2452	1.0469	
3	a_0	191.1798	2.0124	0.9685	174.6026	-20.3929	1.0077	162.5877	-28.2193	1.0210
	a_1	348.9403	-6.5424	0.0104	346.1349	-5.1647	0.0027	343.6743	-0.5769	-0.0056
	a_2	-0.1635	-0.0428		-0.5320	0.7281		0.1836	1.0654	
4	a_0	179.9538	-4.5690	0.9790	160.2093	-24.8253	1.0103	146.4529	-27.7337	1.0154
	a_1	348.6126	-6.6317	0.0082	345.0449	-3.6959	-0.0017	344.0673	1.5741	-0.0099
	a_2	-0.3000	0.1369		-0.3498	0.9302		0.5319	0.9157	
5	a_0	168.2639	-11.0596	0.9872	144.9111	-27.5893	1.0086	131.0550	-25.2489	1.0054
	a_1	348.0077	-6.3608	0.0054	344.3311	-1.8147	-0.0055	345.1629	3.4114	-0.0128
	a_2	-0.4195	0.3415		-0.0191	1.0167		0.6737	0.6744	
6	a_0	155.8508	-17.0743	0.9927	129.2303	-28.3884	1.0031	116.8909	-21.1682	0.9925
	a_1	347.1566	-5.6779	0.0025	344.3031	0.2389	-0.0083	346.5308	4.7560	-0.0143
	a_2	-0.4826	0.5596		0.3298	0.9574		0.6326	0.4281	
7	a_0	142.5258	-22.1884	0.9952	113.8678	-27.1955	0.9947	104.0514	-15.9884	0.9782
	a_1	346.1719	-4.5535	-0.0002	344.9883	2.1649	-0.0101	347.8040	5.6047	-0.0145
	a_2	-0.4365	0.7640		0.5511	0.7870		0.4946	0.2226	
8	a_0	128.2651	-25.9749	0.9950	99.4083	-24.2480	0.9846	92.3465	-10.1644	0.9637
	a_1	345.2788	-3.0133	-0.0025	346.1137	3.7401	-0.0109	348.7940	6.0433	-0.0137
	a_2	-0.2482	0.9107		0.5994	0.5741		0.3271	0.0647	
9	a_0	113.3017	-28.0765	0.9924	86.1199	-19.9379	0.9737	81.4642	-4.0589	0.9500
	a_1	344.7748	-1.1747	-0.0043	347.3255	4.8842	-0.0109	349.4462	6.1684	-0.0123
	a_2	0.0436	0.9557		0.5222	0.3719		0.1634	-0.0570	
10	a_0	98.1259	-28.2970	0.9881	73.9647	-14.6854	0.9627	71.0707	2.0504	0.9377
	a_1	344.8731	0.7520	-0.0056	348.3744	5.6229	-0.0105	349.7699	6.0523	-0.0105
	a_2	0.3254	0.8857		0.3858	0.2016		0.0158	-0.1580	
11	a_0	83.3279	-26.6624	0.9825	62.7217	-8.8638	0.9523	60.8537	7.9428	0.9272
	a_1	345.5451	2.5320	-0.0064	349.1459	6.0220	-0.0097	349.7977	5.7359	-0.0087
	a_2	0.4920	0.7313		0.2343	0.0618		-0.1103	-0.2511	
12	a_0	69.3656	-23.4029	0.9760	52.0989	-2.7825	0.9425	50.5392	13.4256	0.9185
	a_1	346.5476	3.9962	-0.0070	349.6126	6.1432	-0.0088	349.5722	5.2339	-0.0070
	a_2	0.5191	0.5419		0.0891	-0.0570		-0.2077	-0.3446	
13	a_0	56.4307	-18.8688	0.9690	41.7979	3.3014	0.9337	39.9024	18.3130	0.9115
	a_1	347.5962	5.0776	-0.0075	349.7880	6.0281	-0.0080	349.1496	4.5447	-0.0053
	a_2	0.4461	0.3553		-0.0411	-0.1650		-0.2643	-0.4403	
14	a_0	44.4704	-13.4393	0.9615	31.5425	9.1624	0.9257	28.7877	22.4155	0.9062
	a_1	348.4922	5.7849	-0.0079	349.7021	5.6979	-0.0071	348.6117	3.6628	-0.0038
	a_2	0.3241	0.1881		-0.1499	-0.2694		-0.2642	-0.5325	
15	a_0	33.2839	-7.4694	0.9536	21.0930	14.5888	0.9187	17.1366	25.5442	0.9024
	a_1	349.1405	6.1580	-0.0082	349.3969	5.1589	-0.0062	348.0735	2.5941	-0.0024
	a_2	0.1883	0.0418		-0.2282	-0.3732		-0.1974	-0.6084	

$GHA^\circ, DEC^\circ = (a_2x + a_1)x + a_0$ $\quad HP^\circ = a_1x + a_0$ $\quad$ where $x = GMT^h/24$

Day		October			November			December		
		GHA	DEC	HP	GHA	DEC	HP	GHA	DEC	HP
		°	°	°	°	°	°	°	°	°
16	a_0	22.6099	-1.2723	0.9454	10.2610	19.3725	0.9125	5.0153	27.5290	0.9000
	a_1	349.5155	6.2394	-0.0084	348.9329	4.4115	-0.0052	347.6724	1.3715	-0.0010
	a_2	0.0569	-0.0876		-0.2625	-0.4739		-0.0732	-0.6519	
17	a_0	12.1798	4.8771	0.9370	358.9319	23.3082	0.9073	352.6175	28.2486	0.8990
	a_1	349.6267	6.0628	-0.0083	348.3988	3.4610	-0.0040	347.5263	0.0610	0.0004
	a_2	-0.0607	-0.2052		-0.2390	-0.5632		0.0738	-0.6519	
18	a_0	1.7438	10.7323	0.9287	347.0934	26.2046	0.9033	340.2201	27.6585	0.8994
	a_1	349.5015	5.6511	-0.0079	347.9125	2.3300	-0.0025	347.6806	-1.2484	0.0020
	a_2	-0.1569	-0.3148		-0.1542	-0.6282		0.1967	-0.6092	
19	a_0	351.0870	16.0665	0.9208	334.8543	27.9058	0.9008	328.0988	25.8024	0.9014
	a_1	349.1821	5.0202	-0.0070	347.6003	1.0677	-0.0008	348.0840	-2.4700	0.0038
	a_2	-0.2219	-0.4175		-0.0259	-0.6567		0.2614	-0.5361	
20	a_0	340.0467	20.6673	0.9137	322.4314	28.3170	0.8999	316.4442	22.7981	0.9051
	a_1	348.7309	4.1834	-0.0057	347.5511	-0.2516	0.0012	348.6162	-3.5431	0.0057
	a_2	-0.2428	-0.5106		0.1074	-0.6442		0.2591	-0.4489	
21	a_0	328.5356	24.3385	0.9079	310.0920	27.4222	0.9011	305.3183	18.8079	0.9109
	a_1	348.2372	3.1591	-0.0040	347.7737	-1.5442	0.0035	349.1415	-4.4406	0.0079
	a_2	-0.2091	-0.5878		0.2043	-0.5966		0.2001	-0.3597	
22	a_0	316.5654	26.9088	0.9039	298.0707	25.2828	0.9047	294.6579	14.0094	0.9187
	a_1	347.8123	1.9791	-0.0018	348.1918	-2.7394	0.0061	349.5466	-5.1597	0.0101
	a_2	-0.1224	-0.6397		0.2408	-0.5272		0.1000	-0.2729	
23	a_0	304.2578	28.2477	0.9021	286.5028	22.0178	0.9108	284.3018	8.5787	0.9289
	a_1	347.5654	0.6945	0.0007	348.6817	-3.7944	0.0088	349.7501	-5.7061	0.0123
	a_2	-0.0034	-0.6585		0.2154	-0.4488		-0.0293	-0.1847	
24	a_0	291.8222	28.2840	0.9028	275.3985	17.7764	0.9196	274.0194	2.6900	0.9413
	a_1	347.5621	-0.6275	0.0035	349.1188	-4.6921	0.0114	349.6941	-6.0780	0.0142
	a_2	0.1116	-0.6424		0.1400	-0.3679		-0.1812	-0.0851	
25	a_0	279.4975	27.0151	0.9063	264.6549	12.7181	0.9310	263.5289	-3.4705	0.9555
	a_1	347.7929	-1.9158	0.0063	349.4032	-5.4288	0.0137	349.3335	-6.2532	0.0155
	a_2	0.1880	-0.5971		0.0281	-0.2830		-0.3498	0.0409	
26	a_0	267.4788	24.5035	0.9126	254.0833	7.0084	0.9448	252.5089	-9.6793	0.9711
	a_1	348.1775	-3.1120	0.0091	349.4626	-5.9974	0.0155	348.6321	-6.1787	0.0159
	a_2	0.2080	-0.5321		-0.1106	-0.1852		-0.5215	0.2104	
27	a_0	255.8636	20.8611	0.9218	243.4320	0.8285	0.9603	240.6168	-15.6430	0.9871
	a_1	348.6009	-4.1775	0.0116	349.2435	-6.3727	0.0163	347.5795	-5.7656	0.0149
	a_2	0.1723	-0.4549		-0.2691	-0.0605		-0.6585	0.4342	
28	a_0	244.6350	16.2304	0.9334	232.4030	-5.6013	0.9766	227.5374	-20.9689	1.0020
	a_1	348.9509	-5.0890	0.0136	348.7050	-6.5009	0.0157	346.2394	-4.9003	0.0122
	a_2	0.0916	-0.3672		-0.4363	0.1071		-0.6858	0.7003	
29	a_0	233.6752	10.7764	0.9470	220.6687	-11.9905	0.9924	213.0947	-25.1638	1.0143
	a_1	349.1380	-5.8262	0.0147	347.8261	-6.2945	0.0137	344.8341	-3.4908	0.0081
	a_2	-0.0220	-0.2641		-0.5840	0.3285		-0.5124	0.9518	
30	a_0	222.7883	4.6887	0.9618	207.9094	-17.9510	1.0062	197.4242	-27.7000	1.0224
	a_1	349.0965	-6.3587	0.0147	346.6403	-5.6418	0.0100	343.7857	-1.5636	0.0027
	a_2	-0.1587	-0.1360		-0.6523	0.5958		-0.1271	1.0917	
31	a_0	211.7230	-1.8025	0.9765				181.0919	-28.1727	1.0252
	a_1	348.7797	-6.6363	0.0133				343.5403	0.6480	-0.0031
	a_2	-0.3080	0.0280					0.3178	1.0472	

Day		January GHA °	January DEC °	January HP °	February GHA °	February DEC °	February HP °	March GHA °	March DEC °	March HP °
1	a_0	164.9560	-26.4817	1.0220	143.7161	-8.2303	0.9889	164.3407	-4.4861	0.9838
	a_1	344.2085	2.7597	-0.0085	348.2276	6.5592	-0.0154	348.5199	6.7177	-0.0127
	a_2	0.6131	0.8405		0.3318	0.0082		0.1963	-0.0628	
2	a_0	149.7792	-22.8871	1.0135	132.2722	-1.6661	0.9733	153.0539	2.1657	0.9711
	a_1	345.4652	4.4430	-0.0127	348.8898	6.5690	-0.0159	348.9102	6.5872	-0.0139
	a_2	0.6839	0.5655		0.1724	-0.1447		0.0581	-0.2141	
3	a_0	135.9265	-17.8840	1.0008	121.3314	4.7557	0.9574	142.0197	8.5363	0.9571
	a_1	346.8490	5.5673	-0.0153	349.2316	6.2752	-0.0151	349.0227	6.1553	-0.0138
	a_2	0.5939	0.3067		0.0279	-0.2612		-0.0607	-0.3364	
4	a_0	123.3662	-12.0142	0.9854	110.5883	10.7677	0.9422	130.9798	14.3530	0.9432
	a_1	348.0417	6.1721	-0.0163	349.2832	5.7503	-0.0134	348.8959	5.4797	-0.0128
	a_2	0.4340	0.1007		-0.0926	-0.3555		-0.1495	-0.4376	
5	a_0	111.8383	-5.7445	0.9691	99.7771	16.1608	0.9288	119.7253	19.3932	0.9304
	a_1	348.9089	6.3665	-0.0159	349.0923	5.0383	-0.0110	348.5898	4.6022	-0.0109
	a_2	0.2611	-0.0520		-0.1801	-0.4388		-0.1962	-0.5231	
6	a_0	101.0049	0.5676	0.9532	88.6884	20.7587	0.9178	108.1191	23.4709	0.9196
	a_1	349.4283	6.2580	-0.0144	348.7244	4.1598	-0.0083	348.1890	3.5534	-0.0084
	a_2	0.1000	-0.1665		-0.2224	-0.5162		-0.1893	-0.5932	
7	a_0	90.5303	6.6572	0.9387	77.1908	24.4009	0.9095	96.1200	26.4301	0.9112
	a_1	349.6247	5.9230	-0.0124	348.2706	3.1255	-0.0055	347.8025	2.3635	-0.0056
	a_2	-0.0404	-0.2595		-0.2071	-0.5854		-0.1254	-0.6428	
8	a_0	80.1122	12.3189	0.9263	65.2558	26.9400	0.9040	83.7994	28.1504	0.9056
	a_1	349.5393	5.4035	-0.0100	347.8482	1.9511	-0.0028	347.5475	1.0732	-0.0027
	a_2	-0.1539	-0.3446		-0.1311	-0.6369		-0.0179	-0.6642	
9	a_0	69.4960	17.3760	0.9163	52.9754	28.2537	0.9011	71.3315	28.5597	0.9029
	a_1	349.2250	4.7143	-0.0076	347.5819	0.6721	-0.0004	347.5129	-0.2600	0.0002
	a_2	-0.2309	-0.4294		-0.0108	-0.6586		0.1014	-0.6520	
10	a_0	58.4896	21.6590	0.9087	40.5492	28.2675	0.9008	58.9478	27.6486	0.9031
	a_1	348.7545	3.8548	-0.0052	347.5624	-0.6509	0.0017	347.7218	-1.5683	0.0028
	a_2	-0.2570	-0.5139		0.1178	-0.6424		0.1959	-0.6077	
11	a_0	46.9879	24.9984	0.9035	28.2313	26.9753	0.9025	46.8665	25.4740	0.9059
	a_1	348.2307	2.8247	-0.0031	347.8050	-1.9404	0.0035	348.1219	-2.7866	0.0050
	a_2	-0.2189	-0.5898		0.2149	-0.5894		0.2410	-0.5383	
12	a_0	35.0018	27.2322	0.9004	16.2522	24.4472	0.9061	35.2291	22.1509	0.9110
	a_1	347.7849	1.6405	-0.0012	348.2438	-3.1222	0.0050	348.6114	-3.8648	0.0068
	a_2	-0.1173	-0.6423		0.2565	-0.5086		0.2313	-0.4525	
13	a_0	22.6723	28.2301	0.8992	4.7521	20.8184	0.9111	24.0707	17.8356	0.9179
	a_1	347.5478	0.3496	0.0005	348.7646	-4.1409	0.0061	349.0798	-4.7709	0.0081
	a_2	0.0231	-0.6571		0.2399	-0.4118		0.1756	-0.3563	
14	a_0	10.2461	27.9232	0.8997	353.7552	16.2678	0.9171	13.3242	12.7105	0.9260
	a_1	347.5984	-0.9708	0.0019	349.2501	-4.9651	0.0069	349.4349	-5.4847	0.0088
	a_2	0.1584	-0.6281		0.1762	-0.3080		0.0864	-0.2518	
15	a_0	358.0048	26.3255	0.9016	343.1795	10.9969	0.9240	2.8432	6.9764	0.9349
	a_1	347.9242	-2.2316	0.0033	349.6064	-5.5812	0.0075	349.6103	-5.9899	0.0090
	a_2	0.2478	-0.5616		0.0793	-0.2011		-0.0257	-0.1374	

GHA°, DEC° $= (a_2 x + a_1)x + a_0$ HP° $= a_1 x + a_0$ where $x = \mathrm{GMT}^h/24$

Day		January			February			March		
		GHA °	DEC °	HP °	GHA °	DEC °	HP °	GHA °	DEC °	HP °
16	a_0	346.1773	23.5342	0.9048	332.8626	5.2169	0.9316	352.4251	0.8517	0.9440
	a_1	348.4294	-3.3570	0.0045	349.7675	-5.9841	0.0081	349.5603	-6.2669	0.0088
	a_2	0.2726	-0.4720		-0.0406	-0.0906		-0.1530	-0.0095	
17	a_0	334.8786	19.7073	0.9094	322.5867	-0.8553	0.9396	341.8295	-5.4217	0.9527
	a_1	348.9824	-4.3013	0.0059	349.6880	-6.1668	0.0085	349.2543	-6.2887	0.0081
	a_2	0.2369	-0.3736		-0.1763	0.0282		-0.2870	0.1364	
18	a_0	324.0962	15.0345	0.9152	312.0954	-6.9911	0.9482	330.7942	-11.5705	0.9609
	a_1	349.4616	-5.0479	0.0072	349.3357	-6.1131	0.0090	348.6771	-6.0188	0.0073
	a_2	0.1557	-0.2752		-0.3204	0.1627		-0.4126	0.3037	
19	a_0	313.7112	9.7134	0.9225	301.1079	-12.9383	0.9571	319.0570	-17.2817	0.9682
	a_1	349.7766	-5.5981	0.0087	348.6919	-5.7916	0.0093	347.8432	-5.4135	0.0064
	a_2	0.0428	-0.1787		-0.4572	0.3210		-0.4995	0.4906	
20	a_0	303.5277	3.9386	0.9312	289.3406	-18.4050	0.9665	306.4006	-22.2006	0.9746
	a_1	349.8649	-5.9563	0.0102	347.7679	-5.1531	0.0094	346.8277	-4.4310	0.0054
	a_2	-0.0928	-0.0794		-0.5534	0.5066		-0.5006	0.6836	
21	a_0	293.2967	-2.0948	0.9414	276.5552	-23.0473	0.9760	292.7304	-25.9446	0.9800
	a_1	349.6813	-6.1177	0.0117	346.6426	-4.1400	0.0092	345.8058	-3.0565	0.0045
	a_2	-0.2457	0.0322		-0.5537	0.7079		-0.3707	0.8498	
22	a_0	282.7289	-8.1775	0.9531	262.6472	-26.4757	0.9852	278.1707	-28.1494	0.9845
	a_1	349.1901	-6.0579	0.0128	345.5113	-2.7167	0.0083	345.0506	-1.3427	0.0034
	a_2	-0.4081	0.1692		-0.4040	0.8876		-0.1148	0.9429	
23	a_0	271.5078	-14.0625	0.9660	247.7605	-28.3028	0.9935	263.1126	-28.5495	0.9879
	a_1	348.3695	-5.7255	0.0134	344.6881	-0.9253	0.0066	344.8248	0.5597	0.0021
	a_2	-0.5589	0.3448		-0.1110	0.9865		0.1805	0.9281	
24	a_0	259.3165	-19.4387	0.9794	232.3444	-28.2421	1.0001	248.1223	-27.0641	0.9900
	a_1	347.2379	-5.0412	0.0131	344.4721	1.0675	0.0040	345.2041	2.4287	0.0005
	a_2	-0.6501	0.5624		0.2191	0.9587		0.3974	0.8096	
25	a_0	245.9053	-23.9126	0.9926	217.0404	-26.2190	1.0040	233.7255	-23.8295	0.9906
	a_1	345.9118	-3.9154	0.0117	344.9325	2.9997	0.0006	346.0193	4.0539	-0.0014
	a_2	-0.6050	0.7994		0.4491	0.8084		0.4764	0.6256	
26	a_0	231.2171	-27.0248	1.0043	202.4235	-22.4155	1.0046	220.2204	-19.1543	0.9892
	a_1	344.6728	-2.3038	0.0089	345.8535	4.6224	-0.0032	346.9859	5.3056	-0.0036
	a_2	-0.3630	0.9927		0.5200	0.5850		0.4360	0.4190	
27	a_0	215.5348	-28.3346	1.0133	188.7959	-17.2131	1.0014	207.6400	-13.4340	0.9855
	a_1	343.9360	-0.2948	0.0048	346.9075	5.7913	-0.0071	347.8641	6.1412	-0.0060
	a_2	0.0281	1.0590		0.4633	0.3446		0.3267	0.2179	
28	a_0	199.5065	-27.5724	1.0181	176.1641	-11.0817	0.9943	195.8279	-7.0788	0.9796
	a_1	344.0100	1.8466	-0.0002	347.8396	6.4758	-0.0104	348.5187	6.5737	-0.0082
	a_2	0.3968	0.9578		0.3399	0.1237		0.1916	0.0345	
29	a_0	183.9175	-24.7726	1.0179				184.5353	-0.4740	0.9714
	a_1	344.8341	3.7745	-0.0054				348.9003	6.6394	-0.0099
	a_2	0.5922	0.7304					0.0568	-0.1282	
30	a_0	169.3440	-20.2733	1.0125				173.4898	6.0342	0.9614
	a_1	346.0417	5.2355	-0.0100				349.0105	6.3800	-0.0110
	a_2	0.5985	0.4594					-0.0624	-0.2711	
31	a_0	155.9820	-14.5836	1.0024				162.4360	12.1405	0.9503
	a_1	347.2496	6.1479	-0.0135				348.8806	5.8351	-0.0113
	a_2	0.4877	0.2096					-0.1539	-0.3959	

Table 2: MOON, 1987

Day		April			May			June		
		GHA °	DEC °	HP °	GHA °	DEC °	HP °	GHA °	DEC °	HP °
1	a_0	151.1616	17.5776	0.9389	146.3361	27.0826	0.9142	127.4604	25.1960	0.9009
	a_1	348.5657	5.0406	-0.0108	347.4488	2.0051	-0.0059	348.3021	-2.8092	0.0018
	a_2	-0.2043	-0.5029		-0.0644	-0.6786		0.3004	-0.4963	
2	a_0	139.5231	22.1135	0.9281	133.7234	28.4091	0.9083	116.0623	21.8923	0.9027
	a_1	348.1484	4.0314	-0.0095	347.3196	0.6415	-0.0039	348.9111	-3.8014	0.0043
	a_2	-0.2004	-0.5894		0.0770	-0.6772		0.2663	-0.4068	
3	a_0	127.4725	25.5543	0.9186	121.1225	28.3742	0.9044	105.2381	17.6859	0.9070
	a_1	347.7395	2.8481	-0.0075	347.4796	-0.7180	-0.0015	349.4493	-4.6139	0.0070
	a_2	-0.1369	-0.6492		0.1994	-0.6355		0.1852	-0.3222	
4	a_0	115.0773	27.7526	0.9111	108.8030	27.0221	0.9029	94.8703	12.7515	0.9140
	a_1	347.4610	1.5443	-0.0050	347.8878	-1.9918	0.0012	349.8233	-5.2579	0.0097
	a_2	-0.0264	-0.6752		0.2681	-0.5664		0.0727	-0.2421	
5	a_0	102.5145	28.6219	0.9061	96.9589	24.4656	0.9040	84.7634	7.2533	0.9237
	a_1	347.4091	0.1886	-0.0023	348.4329	-3.1252	0.0040	349.9713	-5.7430	0.0122
	a_2	0.0986	-0.6642		0.2725	-0.4851		-0.0620	-0.1593	
6	a_0	90.0243	28.1471	0.9038	85.6633	20.8570	0.9081	74.6695	1.3530	0.9360
	a_1	347.6125	-1.1439	0.0007	348.9848	-4.0952	0.0069	349.8494	-6.0648	0.0142
	a_2	0.1992	-0.6205		0.2218	-0.4022		-0.2141	-0.0618	
7	a_0	77.8372	26.3840	0.9045	74.8681	16.3613	0.9150	64.3015	-4.7709	0.9502
	a_1	348.0195	-2.3874	0.0035	349.4331	-4.8996	0.0095	349.4222	-6.1937	0.0154
	a_2	0.2488	-0.5541		0.1315	-0.3204		-0.3787	0.0660	
8	a_0	66.1054	23.4441	0.9080	64.4303	11.1431	0.9245	53.3416	-10.8950	0.9657
	a_1	348.5251	-3.4968	0.0062	349.6993	-5.5414	0.0117	348.6622	-6.0691	0.0155
	a_2	0.2415	-0.4755		0.0144	-0.2353		-0.5402	0.2393	
9	a_0	54.8708	19.4736	0.9143	54.1411	5.3685	0.9363	41.4613	-16.7201	0.9813
	a_1	349.0140	-4.4484	0.0085	349.7302	-6.0145	0.0133	347.5708	-5.5974	0.0143
	a_2	0.1861	-0.3907		-0.1219	-0.1372		-0.6581	0.4648	
10	a_0	44.0689	14.6365	0.9228	43.7462	-0.7807	0.9496	28.3740	-21.8474	0.9956
	a_1	349.3902	-5.2308	0.0102	349.4876	-6.2934	0.0139	346.2307	-4.6692	0.0116
	a_2	0.0959	-0.2998		-0.2717	-0.0136		-0.6577	0.7240	
11	a_0	33.5526	9.1080	0.9331	32.9589	-7.0844	0.9636	13.9511	-25.7879	1.0073
	a_1	349.5846	-5.8323	0.0113	348.9431	-6.3268	0.0135	344.8835	-3.2105	0.0076
	a_2	-0.0182	-0.1982		-0.4246	0.1477		-0.4600	0.9562	
12	a_0	23.1163	3.0801	0.9444	21.4747	-13.2591	0.9771	358.3824	-28.0398	1.0149
	a_1	349.5500	-6.2317	0.0115	348.0873	-6.0374	0.0118	343.9449	-1.2746	0.0028
	a_2	-0.1485	-0.0783		-0.5545	0.3535		-0.0756	1.0709	
13	a_0	12.5149	-3.2269	0.9559	9.0064	-18.9380	0.9890	342.2601	-28.2447	1.0177
	a_1	349.2534	-6.3924	0.0108	346.9621	-5.3332	0.0091	343.8057	0.8927	-0.0022
	a_2	-0.2873	0.0677		-0.6086	0.5938		0.3367	1.0114	
14	a_0	1.4782	-9.5479	0.9667	355.3618	-23.6726	0.9981	326.4079	-26.3447	1.0154
	a_1	348.6760	-6.2614	0.0094	345.7188	-4.1402	0.0055	344.5107	2.9301	-0.0067
	a_2	-0.4199	0.2447		-0.5158	0.8293		0.5916	0.8091	
15	a_0	349.7323	-15.5603	0.9761	340.5704	-26.9801	1.0036	311.5112	-22.6108	1.0087
	a_1	347.8275	-5.7750	0.0073	344.6632	-2.4651	0.0017	345.7214	4.5497	-0.0101
	a_2	-0.5158	0.4503		-0.2421	0.9911		0.6371	0.5509	

GHA°, DEC° = $(a_2 x + a_1)x + a_0$ HP° = $a_1 x + a_0$ where $x = GMT^h/24$

Day		April GHA °	April DEC °	April HP °	May GHA °	May DEC °	May HP °	June GHA °	June DEC °	June HP °
16	a_0	337.0439	-20.8805	0.9834	324.9993	-28.4536	1.0053	297.8678	-17.5152	0.9986
	a_1	346.7788	-4.8734	0.0050	344.1762	-0.4602	-0.0020	347.0098	5.6458	-0.0122
	a_2	-0.5242	0.6666		0.1338	1.0153		0.5386	0.3086	
17	a_0	323.3013	-25.0834	0.9884	309.3157	-27.9009	1.0032	285.4129	-11.5650	0.9863
	a_1	345.7080	-3.5320	0.0026	344.4648	1.5883	-0.0051	348.0911	6.2557	-0.0131
	a_2	-0.3933	0.8542		0.4464	0.8955		0.3785	0.1109	
18	a_0	308.6215	-27.7591	0.9910	294.2298	-25.4213	0.9981	273.8789	-5.2016	0.9731
	a_1	344.9057	-1.8078	0.0004	345.3853	3.3861	-0.0074	348.8472	6.4718	-0.0130
	a_2	-0.1231	0.9595		0.5846	0.6887		0.2076	-0.0435	
19	a_0	293.4106	-28.6077	0.9914	280.1993	-21.3512	0.9907	262.9304	1.2241	0.9600
	a_1	344.6631	0.1292	-0.0015	346.5739	4.7617	-0.0088	349.2595	6.3813	-0.0123
	a_2	0.1946	0.9455		0.5583	0.4645		0.0489	-0.1679	
20	a_0	278.2732	-27.5354	0.9899	267.3290	-16.1292	0.9818	252.2360	7.4352	0.9478
	a_1	345.0723	2.0332	-0.0030	347.6990	5.6855	-0.0095	349.3533	6.0440	-0.0110
	a_2	0.4293	0.8222		0.4366	0.2638		-0.0871	-0.2765	
21	a_0	263.7766	-24.6839	0.9869	255.4613	-10.1834	0.9723	241.5000	13.2006	0.9367
	a_1	345.9532	3.6823	-0.0042	348.5740	6.2083	-0.0096	349.1734	5.4905	-0.0096
	a_2	0.5145	0.6369		0.2801	0.0967		-0.1908	-0.3793	
22	a_0	250.2435	-20.3688	0.9827	244.3121	-3.8813	0.9626	230.4814	18.3097	0.9271
	a_1	346.9969	4.9552	-0.0052	349.1327	6.3984	-0.0094	348.7836	4.7312	-0.0082
	a_2	0.4712	0.4379		0.1222	-0.0436		-0.2481	-0.4794	
23	a_0	237.7092	-14.9795	0.9775	233.5640	2.4708	0.9532	219.0171	22.5597	0.9189
	a_1	347.9458	5.8280	-0.0061	349.3741	6.3095	-0.0091	348.2772	3.7703	-0.0067
	a_2	0.3551	0.2521		-0.0218	-0.1680		-0.2420	-0.5712	
24	a_0	226.0072	-8.9029	0.9714	222.9138	8.6099	0.9441	207.0540	25.7573	0.9122
	a_1	348.6574	6.3292	-0.0070	349.3263	5.9728	-0.0086	347.7831	2.6234	-0.0054
	a_2	0.2124	0.0860		-0.1421	-0.2855		-0.1639	-0.6415	
25	a_0	214.8740	-2.4908	0.9643	212.0963	14.2947	0.9356	194.6760	27.7386	0.9068
	a_1	349.0807	6.4990	-0.0078	349.0358	5.4012	-0.0080	347.4495	1.3339	-0.0040
	a_2	0.0694	-0.0634		-0.2269	-0.4003		-0.0277	-0.6740	
26	a_0	204.0213	3.9420	0.9565	200.9045	19.2933	0.9276	182.1010	28.3989	0.9028
	a_1	349.2164	6.3707	-0.0085	348.5730	4.5991	-0.0073	347.3955	-0.0209	-0.0026
	a_2	-0.0595	-0.2012		-0.2600	-0.5094		0.1258	-0.6593	
27	a_0	193.1761	10.1088	0.9480	189.2182	23.3811	0.9202	169.6248	27.7199	0.9002
	a_1	349.0928	5.9670	-0.0089	348.0425	3.5768	-0.0065	347.6551	-1.3448	-0.0011
	a_2	-0.1629	-0.3305		-0.2254	-0.6025		0.2473	-0.6013	
28	a_0	182.1046	15.7428	0.9391	177.0375	26.3542	0.9137	157.5285	25.7755	0.8991
	a_1	348.7600	5.3042	-0.0089	347.5828	2.3661	-0.0055	348.1600	-2.5500	0.0006
	a_2	-0.2272	-0.4507		-0.1214	-0.6641		0.3051	-0.5156	
29	a_0	170.6373	20.5941	0.9302	164.5019	28.0558	0.9081	145.9934	22.7119	0.8997
	a_1	348.2966	4.3998	-0.0084	347.3368	1.0306	-0.0042	348.7793	-3.5814	0.0024
	a_2	-0.2363	-0.5560		0.0270	-0.6810		0.2954	-0.4198	
30	a_0	158.6988	24.4362	0.9217	151.8687	28.4060	0.9040	135.0669	18.7126	0.9021
	a_1	347.8144	3.2831	-0.0074	347.3951	-0.3378	-0.0025	349.3766	-4.4200	0.0045
	a_2	-0.1795	-0.6359		0.1728	-0.6503		0.2323	-0.3266	
31	a_0				139.4387	27.4193	0.9014			
	a_1				347.7502	-1.6424	-0.0005			
	a_2				0.2709	-0.5827				

Table 2: MOON, 1987

Day		July GHA	July DEC	July HP	August GHA	August DEC	August HP	September GHA	September DEC	September HP
		°	°	°	°	°	°	°	°	°
1	a_0	124.6737	13.9679	0.9066	114.2166	-7.5733	0.9343	96.3954	-25.9321	0.9713
	a_1	349.8453	-5.0720	0.0068	349.6922	-5.8940	0.0121	346.1219	-3.0400	0.0117
	a_2	0.1329	-0.2404		-0.3608	0.1404		-0.4930	0.8105	
2	a_0	114.6493	8.6572	0.9135	103.5448	-13.3237	0.9465	82.0291	-28.1588	0.9830
	a_1	350.1139	-5.5526	0.0092	348.9683	-5.6185	0.0136	345.1163	-1.4071	0.0110
	a_2	0.0089	-0.1584		-0.5129	0.2886		-0.2585	0.9492	
3	a_0	104.7692	2.9480	0.9227	91.9979	-18.6497	0.9601	66.8934	-28.6160	0.9941
	a_1	350.1341	-5.8707	0.0116	347.9333	-5.0471	0.0146	344.5964	0.5098	0.0093
	a_2	-0.1337	-0.0723		-0.6289	0.4778		0.0608	0.9823	
4	a_0	94.7662	-2.9929	0.9343	79.3020	-23.2145	0.9748	51.5561	-27.1259	1.0035
	a_1	349.8685	-6.0188	0.0137	346.6552	-4.0940	0.0146	344.7346	2.4919	0.0065
	a_2	-0.2923	0.0306		-0.6456	0.6995		0.3325	0.8879	
5	a_0	84.3388	-8.9782	0.9481	65.3148	-26.6047	0.9894	36.6260	-23.7500	1.0101
	a_1	349.2838	-5.9633	0.0153	345.3359	-2.6884	0.0134	345.4230	4.2784	0.0027
	a_2	-0.4597	0.1662		-0.4918	0.9116		0.4617	0.6941	
6	a_0	73.1598	-14.7716	0.9634	50.1656	-28.3789	1.0029	22.5104	-18.7825	1.0128
	a_1	348.3591	-5.6383	0.0160	344.3326	-0.8467	0.0108	346.3640	5.6695	-0.0018
	a_2	-0.6128	0.3488		-0.1628	1.0388		0.4472	0.4529	
7	a_0	60.9044	-20.0563	0.9795	34.3432	-28.1872	1.0137	9.3195	-12.6652	1.0110
	a_1	347.1177	-4.9466	0.0155	344.0133	1.2550	0.0067	347.2671	6.5730	-0.0063
	a_2	-0.6968	0.5808		0.2192	1.0158		0.3428	0.2086	
8	a_0	47.3268	-24.4169	0.9950	18.5812	-25.9199	1.0204	356.9264	-5.8882	1.0046
	a_1	345.6952	-3.7833	0.0134	344.4782	3.3047	0.0017	347.9546	6.9856	-0.0103
	a_2	-0.6267	0.8328		0.4843	0.8432		0.2021	-0.0128	
9	a_0	32.4011	-27.3634	1.0086	3.5455	-21.7773	1.0221	345.0802	1.0806	0.9943
	a_1	344.4116	-2.1023	0.0099	345.4735	4.9975	-0.0038	348.3570	6.9546	-0.0132
	a_2	-0.3436	1.0295		0.5604	0.5845		0.0603	-0.2015	
10	a_0	16.4778	-28.4351	1.0185	349.5780	-16.2009	1.0183	333.4949	7.8305	0.9810
	a_1	343.7171	-0.0170	0.0050	346.6099	6.1639	-0.0088	348.4733	6.5468	-0.0148
	a_2	0.0812	1.0799		0.4901	0.3125		-0.0616	-0.3570	
11	a_0	0.2838	-27.3748	1.0235	336.6751	-9.7295	1.0094	321.9047	14.0177	0.9662
	a_1	343.9026	2.1665	-0.0005	347.5956	6.7823	-0.0127	348.3434	5.8287	-0.0150
	a_2	0.4518	0.9519		0.3484	0.0723		-0.1474	-0.4829	
12	a_0	344.6418	-24.2614	1.0229	324.6158	-2.8789	0.9966	310.1001	19.3614	0.9511
	a_1	344.8389	4.0808	-0.0060	348.2922	6.9198	-0.0152	348.0400	4.8590	-0.0141
	a_2	0.6220	0.7030		0.1893	-0.1215		-0.1813	-0.5821	
13	a_0	330.1024	-19.4834	1.0169	313.0942	3.9161	0.9813	297.9594	23.6368	0.9370
	a_1	346.1055	5.4851	-0.0105	348.6677	6.6710	-0.0161	347.6679	3.6903	-0.0123
	a_2	0.5986	0.4244		0.0412	-0.2727		-0.1518	-0.6533	
14	a_0	316.8038	-13.5790	1.0063	301.8006	10.3119	0.9652	285.4774	26.6731	0.9247
	a_1	347.3122	6.3265	-0.0136	348.7451	6.1217	-0.0156	347.3566	2.3783	-0.0099
	a_2	0.4659	0.1773		-0.0807	-0.3922		-0.0615	-0.6913	
15	a_0	304.5784	-7.0793	0.9926	290.4633	16.0394	0.9496	272.7751	28.3601	0.9149
	a_1	348.2453	6.6733	-0.0152	348.5770	5.3347	-0.0142	347.2313	0.9900	-0.0071
	a_2	0.2969	-0.0181		-0.1633	-0.4903		0.0655	-0.6909	

GHA°, DEC° = $(a_2x + a_1)x + a_0$ HP° = $a_1x + a_0$ where $x = GMT^h/24$

Day		July GHA °	July DEC °	July HP °	August GHA °	August DEC °	August HP °	September GHA °	September DEC °	September HP °
16	a_0	293.1172	-0.4272	0.9774	278.8764	20.8820	0.9354	260.0744	28.6600	0.9077
	a_1	348.8369	6.6312	-0.0154	348.2417	4.3515	-0.0120	347.3661	-0.3968	-0.0043
	a_2	0.1305	-0.1674		-0.1927	-0.5723		0.1886	-0.6528	
17	a_0	282.0815	6.0341	0.9619	266.9262	24.6600	0.9233	247.6308	27.6118	0.9034
	a_1	349.0941	6.2927	-0.0146	347.8468	3.2037	-0.0096	347.7512	-1.7058	-0.0016
	a_2	-0.0162	-0.2853		-0.1583	-0.6355		0.2709	-0.5854	
18	a_0	271.1570	12.0394	0.9473	254.6166	27.2275	0.9137	235.6535	25.3224	0.9018
	a_1	349.0565	5.7204	-0.0130	347.5227	1.9280	-0.0070	348.3016	-2.8781	0.0009
	a_2	-0.1328	-0.3857		-0.0636	-0.6715		0.2959	-0.5008	
19	a_0	260.0792	17.3722	0.9343	242.0784	28.4840	0.9067	224.2504	21.9454	0.9027
	a_1	348.7836	4.9478	-0.0110	347.3935	0.5794	-0.0045	348.9003	-3.8800	0.0031
	a_2	-0.2071	-0.4775		0.0661	-0.6715		0.2673	-0.4093	
20	a_0	248.6553	21.8408	0.9233	229.5405	28.3927	0.9022	213.4166	17.6580	0.9058
	a_1	348.3599	3.9913	-0.0089	347.5298	-0.7689	-0.0022	349.4395	-4.6986	0.0049
	a_2	-0.2242	-0.5611		0.1897	-0.6333		0.1990	-0.3163	
21	a_0	236.7922	25.2696	0.9145	217.2617	26.9920	0.9001	203.0531	12.6451	0.9107
	a_1	347.9015	2.8657	-0.0068	347.9173	-2.0392	-0.0001	349.8404	-5.3313	0.0062
	a_2	-0.1729	-0.6287		0.2707	-0.5636		0.1047	-0.2220	
22	a_0	224.5231	27.5058	0.9077	205.4502	24.3909	0.8999	192.9959	7.0939	0.9170
	a_1	347.5484	1.6030	-0.0048	348.4673	-3.1683	0.0017	350.0516	-5.7762	0.0072
	a_2	-0.0597	-0.6667		0.2933	-0.4745		-0.0064	-0.1236	
23	a_0	212.0147	28.4422	0.9030	194.2102	20.7502	0.9016	183.0384	1.1964	0.9242
	a_1	347.4279	0.2632	-0.0030	349.0608	-4.1176	0.0032	350.0399	-6.0250	0.0078
	a_2	0.0841	-0.6637		0.2620	-0.3774		-0.1288	-0.0160	
24	a_0	199.5293	28.0426	0.9000	183.5316	16.2572	0.9048	172.9468	-4.8421	0.9320
	a_1	347.6017	-1.0698	-0.0013	349.5895	-4.8722	0.0045	349.7825	-6.0594	0.0081
	a_2	0.2125	-0.6181		0.1909	-0.2796		-0.2576	0.1063	
25	a_0	187.3450	26.3563	0.8987	173.3100	11.1075	0.9093	162.4691	-10.7922	0.9402
	a_1	348.0359	-2.3096	0.0003	349.9743	-5.4311	0.0057	349.2655	-5.8498	0.0083
	a_2	0.2886	-0.5399		0.0933	-0.1832		-0.3832	0.2487	
26	a_0	175.6698	23.5088	0.8990	163.3751	5.4951	0.9150	151.3493	-16.3899	0.9484
	a_1	348.6223	-3.3906	0.0018	350.1630	-5.7978	0.0068	348.4930	-5.3552	0.0082
	a_2	0.3006	-0.4446		-0.0221	-0.0863		-0.4841	0.4133	
27	a_0	164.5918	19.6756	0.9008	153.5132	-0.3868	0.9218	139.3574	-21.3280	0.9567
	a_1	349.2303	-4.2795	0.0033	350.1201	-5.9715	0.0079	347.5120	-4.5295	0.0082
	a_2	0.2569	-0.3458		-0.1507	0.0161		-0.5220	0.5939	
28	a_0	154.0773	15.0524	0.9041	143.4798	-6.3398	0.9297	126.3489	-25.2600	0.9648
	a_1	349.7487	-4.9700	0.0049	349.8194	-5.9416	0.0090	346.4493	-3.3381	0.0080
	a_2	0.1737	-0.2509		-0.2881	0.1314		-0.4503	0.7680	
29	a_0	143.9975	9.8333	0.9090	133.0082	-12.1471	0.9386	112.3521	-27.8275	0.9728
	a_1	350.0991	-5.4712	0.0066	349.2416	-5.6822	0.0100	345.5317	-1.7918	0.0076
	a_2	0.0644	-0.1611		-0.4246	0.2682		-0.2493	0.8956	
30	a_0	134.1583	4.2029	0.9156	121.8230	-17.5577	0.9487	97.6401	-28.7229	0.9805
	a_1	350.2301	-5.7936	0.0084	348.3860	-5.1499	0.0110	345.0292	0.0146	0.0069
	a_2	-0.0636	-0.0721		-0.5361	0.4327		0.0299	0.9343	
31	a_0	124.3218	-1.6607	0.9240	109.6721	-22.2710	0.9596			
	a_1	350.1048	-5.9395	0.0103	347.2993	-4.2867	0.0116			
	a_2	-0.2067	0.0246		-0.5778	0.6217				

Table 2: MOON, 1987

Day		October GHA °	October DEC °	October HP °	November GHA °	November DEC °	November HP °	December GHA °	December DEC °	December HP °
1	a_0	82.7044	-27.7753	0.9874	59.2590	-10.5196	0.9852	55.0858	7.5957	0.9628
	a_1	345.1020	1.8980	0.0057	348.2018	6.3928	-0.0034	348.9418	6.2658	-0.0078
	a_2	0.2787	0.8678		0.2161	0.1556		-0.1528	-0.2797	
2	a_0	68.0879	-25.0126	0.9932	47.6738	-3.9748	0.9817	43.8729	13.5790	0.9550
	a_1	345.6797	3.6431	0.0039	348.6336	6.7024	-0.0051	348.6291	5.7057	-0.0080
	a_2	0.4100	0.7167		0.0685	-0.0152		-0.2462	-0.4149	
3	a_0	54.1776	-20.6569	0.9971	36.3730	2.7090	0.9766	32.2551	18.8671	0.9469
	a_1	346.5163	5.0802	0.0015	348.7679	6.6708	-0.0068	348.1261	4.8739	-0.0081
	a_2	0.4116	0.5193		-0.0695	-0.1784		-0.2798	-0.5433	
4	a_0	41.1037	-15.0618	0.9986	25.0690	9.1982	0.9698	20.1025	23.1955	0.9388
	a_1	347.3488	6.1189	-0.0015	348.6235	6.3127	-0.0084	347.5539	3.7827	-0.0081
	a_2	0.3236	0.3082		-0.1813	-0.3352		-0.2326	-0.6506	
5	a_0	28.7735	-8.6390	0.9972	13.5098	15.1726	0.9613	7.4265	26.3263	0.9307
	a_1	347.9993	6.7335	-0.0047	348.2525	5.6401	-0.0096	347.0783	2.4740	-0.0078
	a_2	0.1930	0.1016		-0.2473	-0.4817		-0.1030	-0.7165	
6	a_0	16.9629	-1.8079	0.9924	1.5150	20.3284	0.9517	354.4054	28.0835	0.9228
	a_1	348.3846	6.9342	-0.0078	347.7465	4.6725	-0.0102	346.8694	1.0319	-0.0072
	a_2	0.0537	-0.0917		-0.2457	-0.6075		0.0754	-0.7247	
7	a_0	5.3987	5.0309	0.9845	349.0175	24.3915	0.9415	341.3537	28.3917	0.9156
	a_1	348.4886	6.7478	-0.0104	347.2436	3.4507	-0.0102	347.0267	-0.4251	-0.0061
	a_2	-0.0720	-0.2677		-0.1635	-0.6968		0.2424	-0.6745	
8	a_0	353.8133	11.5077	0.9740	336.1007	27.1446	0.9313	328.6250	27.2940	0.9095
	a_1	348.3384	6.2089	-0.0122	346.9097	2.0486	-0.0094	347.5232	-1.7783	-0.0045
	a_2	-0.1659	-0.4232		-0.0147	-0.7338		0.3457	-0.5832	
9	a_0	341.9848	17.2907	0.9618	322.9991	28.4599	0.9218	316.4945	24.9347	0.9050
	a_1	347.9977	5.3583	-0.0129	346.8818	0.5730	-0.0080	348.2261	-2.9453	-0.0026
	a_2	-0.2090	-0.5537		0.1545	-0.7130		0.3681	-0.4754	
10	a_0	329.7740	22.0931	0.9489	310.0383	28.3213	0.9138	305.0878	21.5161	0.9024
	a_1	347.5692	4.2453	-0.0126	347.1999	-0.8584	-0.0060	348.9706	-3.8946	-0.0002
	a_2	-0.1848	-0.6523		0.2875	-0.6445		0.3225	-0.3708	
11	a_0	317.1603	25.6850	0.9362	297.5270	26.8203	0.9078	294.3791	17.2526	0.9021
	a_1	347.1904	2.9340	-0.0113	347.7863	-2.1495	-0.0035	349.6205	-4.6342	0.0024
	a_2	-0.0911	-0.7096		0.3492	-0.5498		0.2320	-0.2785	
12	a_0	304.2625	27.9091	0.9249	285.6623	24.1230	0.9043	284.2291	12.3416	0.9046
	a_1	347.0047	1.5075	-0.0093	348.4943	-3.2486	-0.0007	350.0872	-5.1897	0.0053
	a_2	0.0488	-0.7189		0.3377	-0.4497		0.1151	-0.1979	
13	a_0	291.3190	28.6985	0.9155	274.4929	20.4266	0.9036	274.4285	6.9556	0.9099
	a_1	347.1061	0.0637	-0.0068	349.1759	-4.1465	0.0022	350.3191	-5.5854	0.0081
	a_2	0.1896	-0.6820		0.2712	-0.3566		-0.0183	-0.1223	
14	a_0	278.6167	28.0815	0.9087	263.9379	15.9253	0.9058	264.7263	1.2497	0.9180
	a_1	347.4941	-1.3041	-0.0040	349.7219	-4.8584	0.0051	350.2841	-5.8317	0.0108
	a_2	0.2872	-0.6111		0.1703	-0.2730		-0.1645	-0.0404	
15	a_0	266.3987	26.1682	0.9046	253.8275	10.7956	0.9110	254.8426	-4.6203	0.9288
	a_1	348.0782	-2.5275	-0.0011	350.0646	-5.4041	0.0078	349.9561	-5.9164	0.0130
	a_2	0.3205	-0.5230		0.0491	-0.1940		-0.3212	0.0617	

GHA°, DEC° $= (a_2x + a_1)x + a_0$ HP° $= a_1x + a_0$ where $x = GMT^h/24$

Day		October GHA °	October DEC °	October HP °	November GHA °	November DEC °	November HP °	December GHA °	December DEC °	December HP °
16	a_0	254.7969	23.1195	0.9035	243.9384	5.1993	0.9188	244.4741	-10.4721	0.9419
	a_1	348.7269	-3.5733	0.0017	350.1643	-5.7935	0.0101	349.3123	-5.7990	0.0146
	a_2	0.2935	-0.4316		-0.0854	-0.1102		-0.4797	0.1991	
17	a_0	243.8159	19.1165	0.9052	234.0143	-0.7023	0.9289	233.3040	-16.0682	0.9565
	a_1	349.3190	-4.4357	0.0043	349.9945	-6.0171	0.0117	348.3458	-5.4078	0.0151
	a_2	0.2226	-0.3432		-0.2298	-0.0095		-0.6131	0.3831	
18	a_0	233.3553	14.3394	0.9096	223.7758	-6.7262	0.9407	221.0356	-21.0883	0.9717
	a_1	349.7671	-5.1220	0.0066	349.5344	-6.0411	0.0126	347.1020	-4.6459	0.0143
	a_2	0.1238	-0.2573		-0.3781	0.1206		-0.6646	0.6094	
19	a_0	223.2438	8.9620	0.9162	212.9293	-12.6433	0.9533	207.4753	-25.1200	0.9860
	a_1	350.0165	-5.6375	0.0083	348.7739	-5.8058	0.0124	345.7449	-3.4229	0.0121
	a_2	0.0080	-0.1683		-0.5130	0.2898		-0.5576	0.8413	
20	a_0	213.2657	3.1582	0.9245	201.1886	-18.1550	0.9658	192.6689	-27.6982	0.9982
	a_1	350.0337	-5.9762	0.0094	347.7361	-5.2307	0.0112	344.6051	-1.7236	0.0087
	a_2	-0.1190	-0.0683		-0.5956	0.4972		-0.2591	1.0044	
21	a_0	203.1776	-2.8838	0.9340	188.3298	-22.8840	0.9771	177.0230	-28.4171	1.0070
	a_1	349.7961	-6.1161	0.0099	346.5236	-4.2352	0.0092	344.0858	0.3092	0.0044
	a_2	-0.2526	0.0514		-0.5642	0.7198		0.1377	1.0238	
22	a_0	192.7183	-8.9454	0.9439	174.2933	-26.3956	0.9862	161.2530	-27.0868	1.0114
	a_1	349.2892	-6.0173	0.0097	345.3705	-2.7848	0.0064	344.3853	2.3764	-0.0001
	a_2	-0.3836	0.1979		-0.3683	0.9048		0.4535	0.8876	
23	a_0	181.6218	-14.7612	0.9537	159.3022	-28.2739	0.9926	146.0946	-23.8275	1.0112
	a_1	348.5160	-5.6252	0.0089	344.6222	-0.9560	0.0033	345.3216	4.1592	-0.0044
	a_2	-0.4906	0.3735		-0.0376	0.9860		0.5790	0.6558	
24	a_0	169.6463	-20.0088	0.9626	143.8936	-28.2451	0.9959	131.9944	-19.0177	1.0068
	a_1	347.5218	-4.8795	0.0077	344.5588	1.0355	0.0003	346.4989	5.4687	-0.0077
	a_2	-0.5339	0.5695		0.2958	0.9300		0.5352	0.4055	
25	a_0	156.6356	-24.3149	0.9703	128.7525	-26.2829	0.9963	119.0256	-13.1482	0.9990
	a_1	346.4342	-3.7364	0.0063	345.1759	2.9069	-0.0022	347.5772	6.2737	-0.0100
	a_2	-0.4632	0.7590		0.4961	0.7650		0.3985	0.1833	
26	a_0	142.6109	-27.2895	0.9766	114.4252	-22.6154	0.9940	106.9978	-6.6950	0.9890
	a_1	345.4892	-2.2070	0.0048	346.1904	4.4390	-0.0043	348.3753	6.6344	-0.0112
	a_2	-0.2550	0.8964		0.5276	0.5533		0.2307	0.0015	
27	a_0	127.8512	-28.5993	0.9814	101.1414	-17.6274	0.9897	95.6005	-0.0623	0.9777
	a_1	344.9749	-0.3978	0.0034	347.2581	5.5425	-0.0057	348.8348	6.6332	-0.0114
	a_2	0.0391	0.9367		0.4390	0.3451		0.0654	-0.1471	
28	a_0	112.8706	-28.0619	0.9848	88.8355	-11.7436	0.9840	84.4976	6.4211	0.9663
	a_1	345.0667	1.4903	0.0020	348.1404	6.2285	-0.0066	348.9616	6.3368	-0.0110
	a_2	0.3027	0.8669		0.2931	0.1614		-0.0808	-0.2755	
29	a_0	98.2430	-25.7078	0.9868	77.2656	-5.3570	0.9774	73.3760	12.4799	0.9552
	a_1	345.6936	3.2324	0.0008	348.7265	6.5482	-0.0071	348.7939	5.7847	-0.0102
	a_2	0.4419	0.7157		0.1322	0.0020		-0.1941	-0.3946	
30	a_0	84.3786	-21.7636	0.9876	66.1212	1.1901	0.9703	61.9746	17.8676	0.9450
	a_1	346.5950	4.6659	-0.0005	348.9886	6.5505	-0.0075	348.3967	4.9944	-0.0093
	a_2	0.4442	0.5281		-0.0212	-0.1421		-0.2565	-0.5081	
31	a_0	71.4160	-16.5736	0.9871				50.1150	22.3518	0.9357
	a_1	347.4932	5.7208	-0.0019				347.8721	3.9755	-0.0083
	a_2	0.3526	0.3369					-0.2472	-0.6100	

Table 2: MOON, 1988

Day		January GHA °	January DEC °	January HP °	February GHA °	February DEC °	February HP °	March GHA °	March DEC °	March HP °
1	a_0	37.7416	25.7159	0.9274	16.9407	26.3237	0.9033	23.7857	20.6224	0.8997
	a_1	347.3660	2.7501	-0.0073	347.9303	-2.3976	-0.0027	349.1578	-4.1378	0.0000
	a_2	-0.1557	-0.6846		0.3344	-0.5350		0.2923	-0.3529	
2	a_0	24.9552	27.7808	0.9202	5.2055	23.3933	0.9006	13.2341	16.1337	0.8997
	a_1	347.0484	1.3730	-0.0063	348.6083	-3.4683	-0.0013	349.7461	-4.8427	0.0016
	a_2	0.0008	-0.7135		0.3389	-0.4285		0.2137	-0.2549	
3	a_0	12.0080	28.4408	0.9139	354.1516	19.4987	0.8993	3.1919	11.0380	0.9013
	a_1	347.0524	-0.0620	-0.0053	349.2924	-4.3244	0.0001	350.1757	-5.3519	0.0030
	a_2	0.1729	-0.6868		0.2892	-0.3222		0.1148	-0.1617	
4	a_0	359.2361	27.6936	0.9086	343.7316	14.8541	0.8994	353.4801	5.5263	0.9042
	a_1	347.4078	-1.4412	-0.0042	349.8748	-4.9673	0.0016	350.4066	-5.6752	0.0042
	a_2	0.3037	-0.6115		0.2041	-0.2233		0.0038	-0.0706	
5	a_0	346.9488	25.6431	0.9044	333.8082	9.6654	0.9010	343.8880	-0.2175	0.9085
	a_1	348.0268	-2.6665	-0.0029	350.2853	-5.4127	0.0032	350.4151	-5.8171	0.0055
	a_2	0.3608	-0.5076		0.0976	-0.1330		-0.1156	0.0231	
6	a_0	335.3361	22.4712	0.9015	324.1886	4.1215	0.9041	334.1849	-6.0094	0.9139
	a_1	348.7577	-3.6814	-0.0013	350.4821	-5.6782	0.0050	350.1843	-5.7725	0.0067
	a_2	0.3452	-0.3969		-0.0226	-0.0477		-0.2404	0.1255	
7	a_0	324.4376	18.3952	0.9001	314.6453	-1.6025	0.9091	324.1261	-11.6540	0.9207
	a_1	349.4541	-4.4733	0.0005	350.4381	-5.7743	0.0070	349.7022	-5.5241	0.0081
	a_2	0.2763	-0.2934		-0.1537	0.0394		-0.3643	0.2434	
8	a_0	314.1658	13.6303	0.9006	304.9268	-7.3354	0.9161	313.4618	-16.9318	0.9288
	a_1	350.0101	-5.0582	0.0027	350.1314	-5.6977	0.0091	348.9691	-5.0404	0.0095
	a_2	0.1745	-0.2024		-0.2934	0.1377		-0.4709	0.3825	
9	a_0	304.3478	8.3714	0.9032	294.7619	-12.8929	0.9251	301.9588	-21.5864	0.9383
	a_1	350.3611	-5.4619	0.0051	349.5436	-5.4262	0.0112	348.0170	-4.2779	0.0109
	a_2	0.0530	-0.1215		-0.4340	0.2580		-0.5287	0.5426	
10	a_0	294.7592	2.7896	0.9083	283.8690	-18.0580	0.9364	289.4478	-25.3182	0.9492
	a_1	350.4689	-5.7051	0.0076	348.6700	-4.9151	0.0132	346.9429	-3.1917	0.0121
	a_2	-0.0824	-0.0434		-0.5541	0.4100		-0.4938	0.7104	
11	a_0	285.1426	-2.9571	0.9159	271.9837	-22.5593	0.9497	275.9002	-27.7965	0.9612
	a_1	350.3056	-5.7939	0.0103	347.5481	-4.0989	0.0148	345.9375	-1.7638	0.0128
	a_2	-0.2301	0.0428		-0.6102	0.5941		-0.3363	0.8532	
12	a_0	275.2148	-8.7061	0.9262	258.9230	-26.0601	0.9645	261.5066	-28.7056	0.9741
	a_1	349.8460	-5.7125	0.0127	346.3056	-2.9091	0.0156	345.2568	-0.0441	0.0129
	a_2	-0.3867	0.1506		-0.5427	0.7896		-0.0829	0.9266	
13	a_0	264.6710	-14.2649	0.9390	244.6906	-28.1764	0.9801	246.6857	-27.8238	0.9870
	a_1	349.0699	-5.4172	0.0148	345.1988	-1.3192	0.0152	345.0991	1.8245	0.0119
	a_2	-0.5388	0.2945		-0.3182	0.9456		0.1736	0.8988	
14	a_0	253.1998	-19.3836	0.9539	229.5778	-28.5489	0.9954	231.9618	-25.1030	0.9990
	a_1	347.9821	-4.8344	0.0160	344.5568	0.5910	0.0133	345.4650	3.6348	0.0097
	a_2	-0.6503	0.4833		0.0076	0.9985		0.3357	0.7719	
15	a_0	240.5316	-23.7302	0.9700	214.1482	-26.9611	1.0088	217.7633	-20.7004	1.0088
	a_1	346.6601	-3.8704	0.0162	344.5878	2.6077	0.0099	346.1545	5.1858	0.0062
	a_2	-0.6551	0.7065		0.2988	0.9145		0.3689	0.5753	

$GHA^\circ, DEC^\circ = (a_2 x + a_1)x + a_0 \qquad HP^\circ = a_1 x + a_0 \qquad$ where $x = GMT^h/24$

Day		January GHA	January DEC	January HP	February GHA	February DEC	February HP	March GHA	March DEC	March HP
		°	°	°	°	°	°	°	°	°
16	a_0	226.5402	-26.8897	0.9862	199.0379	-23.4430	1.0187	204.2853	-14.9441	1.0151
	a_1	345.3213	-2.4500	0.0148	345.2099	4.4495	0.0051	346.9036	6.3389	0.0017
	a_2	-0.4832	0.9177		0.4465	0.7157		0.3001	0.3440	
17	a_0	211.3854	-28.4196	1.0011	184.6943	-18.2831	1.0238	191.4865	-8.2661	1.0168
	a_1	344.3370	-0.5951	0.0119	346.1220	5.8849	-0.0005	347.5084	7.0260	-0.0033
	a_2	-0.1394	1.0373		0.4408	0.4578		0.1764	0.1056	
18	a_0	195.5909	-27.9781	1.0130	171.2549	-11.9458	1.0233	179.1684	-1.1392	1.0135
	a_1	344.0670	1.5038	0.0074	347.0132	6.7981	-0.0061	347.8610	7.2342	-0.0080
	a_2	0.2421	1.0016		0.3360	0.1936		0.0377	-0.1215	
19	a_0	179.9052	-25.4765	1.0205	158.6011	-4.9590	1.0170	167.0644	5.9693	1.0054
	a_1	344.5791	3.5242	0.0021	347.6876	7.1799	-0.0109	347.9323	6.9870	-0.0119
	a_2	0.4925	0.8182		0.1907	-0.0445		-0.0872	-0.3253	
20	a_0	164.9781	-21.1394	1.0225	146.4763	2.1722	1.0061	154.9077	12.6274	0.9934
	a_1	345.5904	5.1658	-0.0035	348.0669	7.0847	-0.0142	347.7504	6.3312	-0.0145
	a_2	0.5510	0.5556		0.0441	-0.2442		-0.1733	-0.4982	
21	a_0	151.1179	-15.4236	1.0190	134.5847	9.0094	0.9917	142.4844	18.4576	0.9788
	a_1	346.7072	6.2737	-0.0083	348.1500	6.5910	-0.0159	347.3932	5.3288	-0.0156
	a_2	0.4674	0.2867		-0.0792	-0.4054		-0.1971	-0.6338	
22	a_0	138.2894	-8.8681	1.0106	122.6537	15.1923	0.9757	129.6815	23.1507	0.9631
	a_1	347.6470	6.8402	-0.0120	347.9837	5.7755	-0.0161	346.9873	4.0539	-0.0153
	a_2	0.3172	0.0527		-0.1595	-0.5334		-0.1429	-0.7243	
23	a_0	126.2501	-1.9791	0.9985	110.4775	20.4323	0.9595	116.5284	26.4795	0.9478
	a_1	348.2808	6.9389	-0.0141	347.6545	4.7041	-0.0151	346.6932	2.5970	-0.0139
	a_2	0.1522	-0.1360		-0.1786	-0.6314		-0.0175	-0.7618	
24	a_0	114.6799	4.8208	0.9844	97.9545	24.5037	0.9444	103.2074	28.3150	0.9338
	a_1	348.5818	6.6621	-0.0148	347.2869	3.4360	-0.0133	346.6575	1.0656	-0.0118
	a_2	0.0002	-0.2860		-0.1256	-0.6966		0.1413	-0.7435	
25	a_0	103.2594	11.1942	0.9695	85.1183	27.2425	0.9310	90.0090	28.6384	0.9220
	a_1	348.5766	6.0868	-0.0144	347.0287	2.0364	-0.0110	346.9470	-0.4270	-0.0093
	a_2	-0.1224	-0.4096		-0.0096	-0.7222		0.2793	-0.6780	
26	a_0	91.7119	16.8692	0.9551	72.1403	28.5571	0.9200	77.2370	27.5354	0.9127
	a_1	348.3237	5.2653	-0.0131	347.0091	0.5856	-0.0086	347.5161	-1.7856	-0.0065
	a_2	-0.1992	-0.5160		0.1344	-0.7032		0.3567	-0.5833	
27	a_0	79.8362	21.6166	0.9420	59.2864	28.4408	0.9114	65.1101	25.1687	0.9062
	a_1	347.9148	4.2306	-0.0114	347.2841	-0.8259	-0.0061	348.2391	-2.9524	-0.0038
	a_2	-0.2123	-0.6066		0.2589	-0.6436		0.3644	-0.4787	
28	a_0	67.5400	25.2392	0.9305	46.8308	26.9731	0.9053	53.7126	21.7397	0.9024
	a_1	347.4793	3.0129	-0.0095	347.8113	-2.1159	-0.0039	348.9744	-3.9087	-0.0012
	a_2	-0.1504	-0.6738		0.3292	-0.5558		0.3164	-0.3768	
29	a_0	54.8716	27.5777	0.9210	34.9715	24.3036	0.9015	43.0017	17.4562	0.9012
	a_1	347.1713	1.6590	-0.0076	348.4784	-3.2282	-0.0018	349.6110	-4.6611	0.0011
	a_2	-0.0232	-0.7044		0.3366	-0.4550		0.2331	-0.2823	
30	a_0	42.0228	28.5326	0.9134				32.8437	12.5146	0.9022
	a_1	347.1248	0.2432	-0.0059				350.0791	-5.2251	0.0030
	a_2	0.1315	-0.6884					0.1305	-0.1935	
31	a_0	29.2818	28.0887	0.9075				23.0509	7.0978	0.9053
	a_1	347.3946	-1.1394	-0.0042				350.3410	-5.6123	0.0046
	a_2	0.2627	-0.6275					0.0174	-0.1054	

Table 2: MOON, 1988

Day		April GHA	April DEC	April HP	May GHA	May DEC	May HP	June GHA	June DEC	June HP
		°	°	°	°	°	°	°	°	°
1	a_0	13.4068	1.3820	0.9099	13.1890	-14.0463	0.9325	353.5241	-27.8655	0.9737
	a_1	350.3764	-5.8243	0.0058	349.1295	-5.4228	0.0091	345.3316	-1.3726	0.0081
	a_2	-0.1017	-0.0118		-0.4509	0.3143		-0.1930	0.9134	
2	a_0	3.6789	-4.4519	0.9157	1.8662	-19.1514	0.9417	338.6692	-28.3245	0.9817
	a_1	350.1729	-5.8500	0.0067	348.2177	-4.7969	0.0090	344.9471	0.4722	0.0058
	a_2	-0.2239	0.0937		-0.5155	0.4850		0.1211	0.9223	
3	a_0	353.6254	-10.2057	0.9225	349.5690	-23.4596	0.9507	323.7424	-26.9321	0.9875
	a_1	349.7236	-5.6654	0.0074	347.1703	-3.8257	0.0085	345.2080	2.3312	0.0034
	a_2	-0.3427	0.2166		-0.4898	0.6632		0.3641	0.8164	
4	a_0	343.0043	-15.6514	0.9299	336.2526	-26.6190	0.9591	309.3166	-23.7883	0.9909
	a_1	349.0336	-5.2349	0.0080	346.1724	-2.4917	0.0077	345.9588	3.9704	0.0010
	a_2	-0.4423	0.3601		-0.3405	0.8132		0.4602	0.6369	
5	a_0	331.5945	-20.5229	0.9379	322.0898	-28.2960	0.9668	295.7349	-19.1851	0.9919
	a_1	348.1391	-4.5162	0.0084	345.4819	-0.8518	0.0067	346.8953	5.2443	-0.0010
	a_2	-0.4938	0.5206		-0.0882	0.8909		0.4217	0.4345	
6	a_0	319.2406	-24.5151	0.9462	307.4890	-28.2573	0.9735	283.0494	-13.5103	0.9909
	a_1	347.1362	-3.4731	0.0087	345.3122	0.9447	0.0057	347.7464	6.1106	-0.0026
	a_2	-0.4581	0.6828		0.1775	0.8695		0.3015	0.2408	
7	a_0	305.9217	-27.3027	0.9550	292.9823	-26.4453	0.9792	271.0941	-7.1626	0.9882
	a_1	346.2038	-2.1004	0.0090	345.6854	2.6943	0.0046	348.3516	6.5894	-0.0040
	a_2	-0.3103	0.8161		0.3549	0.7592		0.1470	0.0650	
8	a_0	291.8201	-28.5856	0.9640	279.0237	-22.9951	0.9838	259.5894	-0.5116	0.9843
	a_1	345.5759	-0.4561	0.0091	346.4139	4.2173	0.0035	348.6446	6.7177	-0.0051
	a_2	-0.0757	0.8833		0.4021	0.5963		-0.0122	-0.0969	
9	a_0	277.3252	-28.1589	0.9732	265.8386	-18.1852	0.9873	248.2189	6.1059	0.9792
	a_1	345.4316	1.3238	0.0089	347.2306	5.4104	0.0022	348.6166	6.5233	-0.0060
	a_2	0.1626	0.8617		0.3414	0.4154		-0.1565	-0.2531	
10	a_0	262.9226	-25.9755	0.9821	253.4081	-12.3632	0.9896	236.6767	12.3730	0.9732
	a_1	345.7737	3.0574	0.0082	347.9194	6.2404	0.0007	348.2965	6.0168	-0.0069
	a_2	0.3169	0.7575		0.2178	0.2347		-0.2655	-0.4081	
11	a_0	249.0140	-22.1638	0.9904	241.5421	-5.8918	0.9903	224.7067	17.9786	0.9663
	a_1	346.4244	4.5782	0.0068	348.3563	6.7092	-0.0011	347.7539	5.1988	-0.0077
	a_2	0.3528	0.5975		0.0687	0.0577		-0.3116	-0.5580	
12	a_0	235.7900	-16.9921	0.9972	229.9640	0.8714	0.9893	212.1500	22.6168	0.9586
	a_1	347.1412	5.7757	0.0046	348.4919	6.8246	-0.0031	347.1158	4.0777	-0.0083
	a_2	0.2907	0.4088		-0.0806	-0.1186		-0.2659	-0.6866	
13	a_0	223.2195	-10.8118	1.0018	218.3726	7.5737	0.9861	199.0029	26.0062	0.9502
	a_1	347.7280	6.5940	0.0015	348.3258	6.5873	-0.0054	346.5710	2.6956	-0.0088
	a_2	0.1715	0.2079		-0.2091	-0.2964		-0.1204	-0.7678	
14	a_0	211.1161	-4.0141	1.0034	206.4876	13.8609	0.9808	185.4578	27.9337	0.9414
	a_1	348.0719	7.0097	-0.0020	347.8986	5.9931	-0.0075	346.3263	1.1488	-0.0089
	a_2	0.0306	0.0028		-0.2913	-0.4716		0.0879	-0.7792	
15	a_0	199.2157	2.9942	1.0014	194.0948	19.3791	0.9732	171.8761	28.3045	0.9325
	a_1	348.1304	7.0144	-0.0057	347.3024	5.0456	-0.0094	346.5098	-0.4190	-0.0085
	a_2	-0.1048	-0.2008		-0.2962	-0.6300		0.2850	-0.7181	

GHA°, DEC° = $(a_2x + a_1)x + a_0$ HP° = $a_1x + a_0$ where $x = GMT^h/24$

Day		April			May			June		
		GHA °	DEC °	HP °	GHA °	DEC °	HP °	GHA °	DEC °	HP °
16	a_0	187.2391	9.8037	0.9956	181.1030	23.7922	0.9638	158.6734	27.1697	0.9240
	a_1	347.9142	6.6104	-0.0091	346.6951	3.7774	-0.0106	347.0938	-1.8602	-0.0077
	a_2	-0.2101	-0.3955		-0.2008	-0.7476		0.4061	-0.6063	
17	a_0	174.9422	16.0150	0.9864	167.6011	26.8208	0.9532	146.1740	24.7059	0.9163
	a_1	347.4834	5.8150	-0.0117	346.2841	2.2709	-0.0111	347.9193	-3.0732	-0.0063
	a_2	-0.2573	-0.5690		-0.0172	-0.7989		0.4318	-0.4758	
18	a_0	162.1690	21.2583	0.9746	153.8724	28.2933	0.9420	134.5241	21.1594	0.9100
	a_1	346.9551	4.6698	-0.0133	346.2517	0.6621	-0.0108	348.7918	-4.0225	-0.0044
	a_2	-0.2203	-0.7041		0.1953	-0.7726		0.3813	-0.3515	
19	a_0	148.9065	25.2224	0.9613	140.3228	28.1847	0.9312	123.6952	16.7877	0.9055
	a_1	346.5028	3.2518	-0.0136	346.6541	-0.8902	-0.0097	349.5589	-4.7225	-0.0022
	a_2	-0.0926	-0.7816		0.3601	-0.6823		0.2845	-0.2451	
20	a_0	135.3204	27.6924	0.9476	127.3386	26.6148	0.9214	113.5361	11.8220	0.9034
	a_1	346.3137	1.6782	-0.0129	347.3887	-2.2570	-0.0080	350.1300	-5.2103	0.0004
	a_2	0.0926	-0.7888		0.4328	-0.5591		0.1643	-0.1567	
21	a_0	121.7303	28.5830	0.9346	115.1598	23.8013	0.9134	103.8276	6.4567	0.9038
	a_1	346.5051	0.0925	-0.0113	348.2656	-3.3738	-0.0057	350.4595	-5.5225	0.0032
	a_2	0.2704	-0.7304		0.4163	-0.4331		0.0325	-0.0797	
22	a_0	108.5082	27.9471	0.9233	103.8401	19.9966	0.9077	94.3166	0.8560	0.9070
	a_1	347.0583	-1.3725	-0.0091	349.1048	-4.2373	-0.0030	350.5253	-5.6825	0.0060
	a_2	0.3829	-0.6289		0.3387	-0.3214		-0.1068	-0.0041	
23	a_0	95.9500	25.9482	0.9142	93.2813	15.4399	0.9047	84.7321	-4.8288	0.9131
	a_1	347.8361	-2.6305	-0.0064	349.7852	-4.8776	-0.0002	350.3121	-5.6930	0.0087
	a_2	0.4102	-0.5119		0.2280	-0.2279		-0.2523	0.0824	
24	a_0	84.1955	22.8081	0.9077	83.2918	10.3361	0.9045	74.7889	-10.4371	0.9218
	a_1	348.6648	-3.6526	-0.0036	350.2425	-5.3319	0.0027	349.8067	-5.5325	0.0111
	a_2	0.3669	-0.4000		0.1017	-0.1476		-0.3987	0.1926	
25	a_0	73.2254	18.7576	0.9041	73.6332	4.8581	0.9071	64.1942	-15.7741	0.9329
	a_1	349.4032	-4.4503	-0.0008	350.4465	-5.6272	0.0054	349.0046	-5.1528	0.0128
	a_2	0.2793	-0.3013		-0.0325	-0.0712		-0.5288	0.3376	
26	a_0	62.9055	14.0078	0.9034	64.0443	-0.8385	0.9125	52.6685	-20.5856	0.9458
	a_1	349.9639	-5.0514	0.0019	350.3820	-5.7714	0.0078	347.9349	-4.4823	0.0137
	a_2	0.1687	-0.2147		-0.1720	0.0127		-0.6041	0.5197	
27	a_0	53.0355	8.7434	0.9053	54.2513	-6.5950	0.9204	40.0002	-24.5440	0.9596
	a_1	350.3022	-5.4806	0.0043	350.0377	-5.7495	0.0097	346.7057	-3.4426	0.0136
	a_2	0.0471	-0.1336		-0.3138	0.1159		-0.5654	0.7205	
28	a_0	43.3821	3.1310	0.9096	43.9724	-12.2259	0.9301	26.1446	-27.2626	0.9733
	a_1	350.3968	-5.7488	0.0063	349.4074	-5.5226	0.0110	345.5516	-1.9923	0.0123
	a_2	-0.0804	-0.0489		-0.4474	0.2490		-0.3684	0.8912	
29	a_0	33.6957	-2.6646	0.9160	32.9303	-17.4959	0.9411	11.3343	-28.3622	0.9857
	a_1	350.2361	-5.8490	0.0078	348.5045	-5.0292	0.0115	344.8049	-0.1917	0.0099
	a_2	-0.2110	0.0488		-0.5465	0.4168		-0.0476	0.9670	
30	a_0	23.7180	-8.4624	0.9238	20.8881	-22.1044	0.9526	356.0981	-27.5881	0.9956
	a_1	349.8127	-5.7550	0.0087	347.3956	-4.1972	0.0111	344.7219	1.7616	0.0066
	a_2	-0.3395	0.1680		-0.5641	0.6095		0.2682	0.9096	
31	a_0				7.7219	-25.6883	0.9637			
	a_1				346.2455	-2.9730	0.0099			
	a_2				-0.4486	0.7934				

Table 2: MOON, 1988

Day		July GHA °	July DEC °	July HP °	August GHA °	August DEC °	August HP °	September GHA °	September DEC °	September HP °
1	a_0	341.0920	-24.9203	1.0022	320.0525	-3.9559	1.0107	302.6709	19.5621	0.9818
	a_1	345.2828	3.5932	0.0028	347.8904	7.0700	-0.0092	347.1506	5.0886	-0.0155
	a_2	0.4528	0.7394		0.1423	-0.0577		-0.2033	-0.6682	
2	a_0	326.8282	-20.5924	1.0051	308.0820	3.0525	1.0014	289.6196	23.9807	0.9663
	a_1	346.2097	5.0754	-0.0010	348.1728	6.9499	-0.0117	346.7316	3.7447	-0.0149
	a_2	0.4756	0.5142		-0.0089	-0.2432		-0.1328	-0.7519	
3	a_0	313.5116	-15.0075	1.0041	296.2433	9.7559	0.9896	276.2213	26.9729	0.9514
	a_1	347.1728	6.1013	-0.0043	348.1497	6.4596	-0.0129	346.4579	2.2323	-0.0134
	a_2	0.3832	0.2861		-0.1357	-0.4010		0.0107	-0.7808	
4	a_0	301.0646	-8.6243	0.9998	284.2555	15.8118	0.9766	262.6934	28.4251	0.9379
	a_1	347.9436	6.6691	-0.0068	347.8696	5.6544	-0.0131	346.4800	0.6625	-0.0115
	a_2	0.2355	0.0814		-0.2158	-0.5359		0.1818	-0.7510	
5	a_0	289.2403	-1.8775	0.9929	271.9091	20.9279	0.9635	249.3581	28.3382	0.9264
	a_1	348.4144	6.8280	-0.0085	347.4262	4.5785	-0.0125	346.8520	-0.8452	-0.0094
	a_2	0.0736	-0.0954		-0.2256	-0.6474		0.3215	-0.6718	
6	a_0	277.7252	4.8519	0.9844	259.1113	24.8574	0.9510	236.5331	26.8234	0.9169
	a_1	348.5586	6.6346	-0.0094	346.9626	3.2777	-0.0113	347.5063	-2.1914	-0.0073
	a_2	-0.0783	-0.2511		-0.1502	-0.7257		0.3918	-0.5631	
7	a_0	266.2030	11.2324	0.9750	245.9268	27.4088	0.9397	224.4311	24.0714	0.9097
	a_1	348.3961	6.1308	-0.0097	346.6544	1.8183	-0.0100	348.2993	-3.3175	-0.0053
	a_2	-0.2009	-0.3943		-0.0015	-0.7562		0.3894	-0.4453	
8	a_0	254.3967	16.9661	0.9654	232.5833	28.4715	0.9297	213.1186	20.3109	0.9044
	a_1	347.9846	5.3405	-0.0095	346.6522	0.2976	-0.0085	349.0841	-4.2067	-0.0034
	a_2	-0.2715	-0.5285		0.1738	-0.7299		0.3329	-0.3322	
9	a_0	242.1100	21.7757	0.9558	219.4122	28.0408	0.9212	202.5338	15.7742	0.9010
	a_1	347.4284	4.2800	-0.0092	347.0085	-1.1684	-0.0071	349.7531	-4.8694	-0.0017
	a_2	-0.2637	-0.6471		0.3160	-0.6522		0.2448	-0.2289	
10	a_0	229.2769	25.4069	0.9466	206.7382	26.2225	0.9141	192.5296	10.6779	0.8993
	a_1	346.8877	2.9792	-0.0087	347.6520	-2.4757	-0.0057	350.2441	-5.3259	-0.0002
	a_2	-0.1622	-0.7324		0.3868	-0.5420		0.1409	-0.1351	
11	a_0	216.0061	27.6531	0.9379	194.7770	23.2073	0.9084	182.9122	5.2187	0.8991
	a_1	346.5561	1.5052	-0.0081	348.4354	-3.5597	-0.0044	350.5265	-5.5954	0.0013
	a_2	0.0143	-0.7628		0.3833	-0.4209		0.0297	-0.0470	
12	a_0	202.5805	28.3962	0.9298	183.5944	19.2291	0.9041	173.4660	-0.4219	0.9004
	a_1	346.5878	-0.0298	-0.0075	349.2082	-4.4000	-0.0030	350.5861	-5.6896	0.0027
	a_2	0.2079	-0.7277		0.3246	-0.3046		-0.0854	0.0403	
13	a_0	189.3791	27.6406	0.9223	173.1253	14.5267	0.9011	163.9642	-6.0693	0.9031
	a_1	347.0147	-1.4916	-0.0067	349.8607	-5.0071	-0.0016	350.4152	-5.6101	0.0043
	a_2	0.3519	-0.6368		0.2331	-0.1999		-0.2023	0.1326	
14	a_0	176.7471	25.5148	0.9155	163.2169	9.3215	0.8995	154.1748	-11.5447	0.9074
	a_1	347.7315	-2.7674	-0.0058	350.3286	-5.4053	0.0000	350.0095	-5.3468	0.0059
	a_2	0.4114	-0.5150		0.1246	-0.1074		-0.3162	0.2359	
15	a_0	164.8896	22.2350	0.9097	153.6676	3.8105	0.8995	143.8660	-16.6531	0.9133
	a_1	348.5643	-3.7965	-0.0045	350.5787	-5.6193	0.0018	349.3734	-4.8774	0.0077
	a_2	0.3912	-0.3882		0.0069	-0.0231		-0.4152	0.3554	

$GHA^\circ, DEC^\circ = (a_2 x + a_1)x + a_0 \quad HP^\circ = a_1 x + a_0 \quad$ where $x = GMT^h/24$

Day		July GHA °	July DEC °	July HP °	August GHA °	August DEC °	August HP °	September GHA °	September DEC °	September HP °
16	a_0	153.8435	18.0527	0.9052	144.2506	-1.8301	0.9013	132.8229	-21.1722	0.9210
	a_1	349.3525	-4.5705	-0.0030	350.5932	-5.6657	0.0038	348.5349	-4.1686	0.0096
	a_2	0.3163	-0.2724		-0.1170	0.0593		-0.4758	0.4922	
17	a_0	143.5101	13.2119	0.9022	134.7241	-7.4346	0.9052	120.8822	-24.8456	0.9305
	a_1	349.9880	-5.1127	-0.0011	350.3594	-5.5485	0.0061	347.5703	-3.1842	0.0114
	a_2	0.2110	-0.1732		-0.2455	0.1481		-0.4646	0.6379	
18	a_0	133.7065	7.9276	0.9011	124.8353	-12.8329	0.9112	107.9903	-27.3892	0.9420
	a_1	350.4114	-5.4575	0.0011	349.8673	-5.2551	0.0084	346.6263	-1.9039	0.0131
	a_2	0.0902	-0.0884		-0.3730	0.2524		-0.3546	0.7705	
19	a_0	124.2053	2.3834	0.9022	114.3273	-17.8329	0.9197	94.2660	-28.5208	0.9551
	a_1	350.5926	-5.6336	0.0036	349.1168	-4.7540	0.0108	345.9079	-0.3534	0.0143
	a_2	-0.0397	-0.0109		-0.4839	0.3805		-0.1564	0.8563	
20	a_0	114.7554	-3.2595	0.9057	102.9589	-22.2032	0.9305	80.0220	-28.0178	0.9694
	a_1	350.5140	-5.6565	0.0062	348.1387	-3.9962	0.0131	345.5981	1.3717	0.0147
	a_2	-0.1769	0.0687		-0.5465	0.5347		0.0664	0.8654	
21	a_0	105.0895	-8.8454	0.9120	90.5517	-25.6612	0.9436	65.6899	-25.7822	0.9842
	a_1	350.1604	-5.5219	0.0089	347.0287	-2.9266	0.0149	345.7445	3.1145	0.0140
	a_2	-0.3190	0.1618		-0.5156	0.7029		0.2306	0.7885	
22	a_0	94.9280	-14.2030	0.9209	77.0681	-27.8818	0.9585	51.6662	-21.8824	0.9982
	a_1	349.5203	-5.2026	0.0115	345.9791	-1.5142	0.0159	346.2220	4.7002	0.0119
	a_2	-0.4571	0.2805		-0.3584	0.8512		0.2890	0.6388	
23	a_0	83.9891	-19.1219	0.9324	62.6941	-28.5433	0.9745	38.1764	-16.5474	1.0102
	a_1	348.5990	-4.6466	0.0136	345.2534	0.2019	0.0158	346.8123	5.9833	0.0083
	a_2	-0.5660	0.4333		-0.1006	0.9304		0.2476	0.4392	
24	a_0	72.0214	-23.3313	0.9460	47.8523	-27.4116	0.9903	25.2341	-10.1297	1.0185
	a_1	347.4515	-3.7830	0.0151	345.0603	2.0794	0.0142	347.3140	6.8642	0.0035
	a_2	-0.5990	0.6170		0.1625	0.9026		0.1409	0.2107	
25	a_0	58.8760	-26.4935	0.9612	33.0786	-24.4324	1.0046	12.6862	-3.0598	1.0220
	a_1	346.2311	-2.5459	0.0156	345.4046	3.8985	0.0111	347.5971	7.2857	-0.0019
	a_2	-0.4996	0.8043		0.3291	0.7667		0.0065	-0.0290	
26	a_0	44.6127	-28.2324	0.9768	18.8130	-19.7715	1.0158	0.2872	4.1921	1.0201
	a_1	345.2132	-0.9248	0.0147	346.0813	5.4397	0.0066	347.6070	7.2256	-0.0072
	a_2	-0.2508	0.9398		0.3622	0.5544		-0.1229	-0.2638	
27	a_0	29.5816	-28.2169	0.9916	5.2551	-13.7825	1.0224	347.7693	11.1495	1.0128
	a_1	344.7108	0.9740	0.0124	346.8175	6.5506	0.0011	347.3533	6.6936	-0.0116
	a_2	0.0719	0.9643		0.2903	0.3070		-0.2160	-0.4779	
28	a_0	14.3698	-26.2809	1.0041	352.3602	-6.9301	1.0235	334.9061	17.3615	1.0011
	a_1	344.8731	2.9203	0.0088	347.4026	7.1626	-0.0044	346.9089	5.7309	-0.0145
	a_2	0.3328	0.8573		0.1623	0.0577		-0.2400	-0.6549	
29	a_0	359.5780	-22.5075	1.0129	339.9222	0.2854	1.0190	321.5765	22.4350	0.9865
	a_1	345.5625	4.6450	0.0042	347.7267	7.2739	-0.0093	346.4146	4.4117	-0.0159
	a_2	0.4439	0.6517		0.0198	-0.1724		-0.1701	-0.7769	
30	a_0	345.5839	-17.2159	1.0171	327.6660	7.3828	1.0097	307.8242	26.0687	0.9705
	a_1	346.4674	5.9504	-0.0008	347.7617	6.9242	-0.0128	346.0642	2.8467	-0.0159
	a_2	0.4124	0.4040		-0.1063	-0.3728		-0.0123	-0.8279	
31	a_0	332.4612	-10.8665	1.0162	315.3197	13.9308	0.9967			
	a_1	347.3002	6.7553	-0.0055	347.5407	6.1733	-0.0149			
	a_2	0.2942	0.1598		-0.1892	-0.5393				

Day		October			November			December		
		GHA °	DEC °	HP °	GHA °	DEC °	HP °	GHA °	DEC °	HP °
1	a_0	293.8801	28.0880	0.9545	273.3078	22.3384	0.9182	272.2245	9.6306	0.9060
	a_1	346.0395	1.1804	-0.0147	348.4426	-3.7870	-0.0081	350.2366	-5.3484	-0.0023
	a_2	0.1839	-0.8033		0.4337	-0.3804		0.1482	-0.1043	
2	a_0	280.1069	28.4669	0.9398	262.1820	18.1734	0.9101	262.6064	4.1794	0.9037
	a_1	346.4169	-0.4332	-0.0128	349.3141	-4.5443	-0.0053	350.5329	-5.5562	0.0007
	a_2	0.3476	-0.7171		0.3313	-0.2682		0.0118	-0.0295	
3	a_0	266.8732	27.3190	0.9270	251.8248	13.3628	0.9048	253.1482	-1.4047	0.9044
	a_1	347.1253	-1.8700	-0.0104	349.9779	-5.0783	-0.0025	350.5562	-5.6162	0.0034
	a_2	0.4309	-0.5970		0.2097	-0.1758		-0.1244	0.0462	
4	a_0	254.4294	24.8546	0.9166	242.0098	8.1104	0.9023	243.5772	-6.9728	0.9078
	a_1	347.9981	-3.0633	-0.0078	350.3975	-5.4288	0.0001	350.3067	-5.5262	0.0059
	a_2	0.4299	-0.4703		0.0828	-0.0951		-0.2583	0.1343	
5	a_0	242.8562	21.3234	0.9088	232.4874	2.5882	0.9024	233.6230	-12.3625	0.9138
	a_1	348.8646	-4.0018	-0.0052	350.5628	-5.6193	0.0024	349.7874	-5.2613	0.0079
	a_2	0.3679	-0.3534		-0.0444	-0.0164		-0.3827	0.2445	
6	a_0	232.0866	16.9703	0.9036	223.0031	-3.0456	0.9049	223.0257	-17.3764	0.9217
	a_1	349.6034	-4.7065	-0.0028	350.4734	-5.6536	0.0044	349.0156	-4.7762	0.0093
	a_2	0.2722	-0.2510		-0.1700	0.0699		-0.4779	0.3821	
7	a_0	221.9599	12.0147	0.9008	213.3039	-8.6272	0.9093	211.5628	-21.7672	0.9310
	a_1	350.1489	-5.2071	-0.0007	350.1319	-5.5165	0.0059	348.0472	-4.0139	0.0100
	a_2	0.1617	-0.1603		-0.2901	0.1719		-0.5093	0.5421	
8	a_0	212.2681	6.6490	0.9001	203.1435	-13.9693	0.9152	199.1022	-25.2356	0.9410
	a_1	350.4726	-5.5274	0.0012	349.5479	-5.1759	0.0070	347.0111	-2.9268	0.0100
	a_2	0.0462	-0.0754		-0.3939	0.2949		-0.4359	0.7019	
9	a_0	202.7844	1.0481	0.9013	192.2962	-18.8473	0.9222	185.6814	-27.4582	0.9510
	a_1	350.5649	-5.6786	0.0027	348.7521	-4.5884	0.0077	346.1238	-1.5136	0.0094
	a_2	-0.0705	0.0107		-0.4589	0.4383		-0.2436	0.8209	
10	a_0	193.2764	-4.6178	0.9040	180.5895	-22.9942	0.9299	171.5669	-28.1501	0.9604
	a_1	350.4234	-5.6586	0.0041	347.8213	-3.7114	0.0080	345.6336	0.1423	0.0082
	a_2	-0.1863	0.1044		-0.4523	0.5906		0.0180	0.8578	
11	a_0	183.5111	-10.1698	0.9081	167.9608	-26.1123	0.9379	157.2233	-27.1512	0.9686
	a_1	350.0492	-5.4519	0.0052	346.9016	-2.5252	0.0081	345.6817	1.8713	0.0067
	a_2	-0.2964	0.2109		-0.3460	0.7262		0.2492	0.7976	
12	a_0	173.2620	-15.4082	0.9133	154.5205	-27.9095	0.9460	143.1568	-24.4851	0.9754
	a_1	349.4525	-5.0323	0.0063	346.1996	-1.0628	0.0080	346.1990	3.4745	0.0051
	a_2	-0.3894	0.3334		-0.1478	0.8109		0.3704	0.6621	
13	a_0	162.3239	-20.1042	0.9196	140.5771	-28.1613	0.9540	129.7262	-20.3521	0.9805
	a_1	348.6659	-4.3668	0.0073	345.9063	0.5711	0.0079	346.9559	4.8011	0.0035
	a_2	-0.4435	0.4703		0.0808	0.8188		0.3686	0.4900	
14	a_0	150.5467	-23.9979	0.9270	126.5678	-26.7728	0.9619	117.0488	-15.0647	0.9840
	a_1	347.7668	-3.4254	0.0084	346.0812	2.2186	0.0077	347.7027	5.7802	0.0021
	a_2	-0.4290	0.6110		0.2556	0.7492		0.2775	0.3120	
15	a_0	137.8866	-26.8097	0.9353	112.9062	-23.8075	0.9697	105.0262	-8.9761	0.9860
	a_1	346.8952	-2.1986	0.0094	346.6090	3.7227	0.0074	348.2620	6.4027	0.0008
	a_2	-0.3239	0.7342		0.3258	0.6249		0.1383	0.1412	

$GHA^\circ, DEC^\circ = (a_2x + a_1)x + a_0$ $HP^\circ = a_1x + a_0$ where $x = GMT^h/24$

Day		October GHA °	October DEC °	October HP °	November GHA °	November DEC °	November HP °	December GHA °	December DEC °	December HP °
16	a_0	124.4618	-28.2726	0.9448	99.8404	-19.4630	0.9771	93.4232	-2.4357	0.9868
	a_1	346.2393	-0.7214	0.0104	347.2735	4.9747	0.0069	348.5393	6.6843	-0.0004
	a_2	-0.1397	0.8111		0.2936	0.4734		-0.0190	-0.0233	
17	a_0	110.5656	-28.1827	0.9552	87.4053	-14.0182	0.9840	81.9404	4.2219	0.9864
	a_1	345.9627	0.9115	0.0112	347.8682	5.9223	0.0060	348.4992	6.6382	-0.0016
	a_2	0.0660	0.8197		0.1904	0.3116		-0.1728	-0.1886	
18	a_0	96.5975	-26.4529	0.9665	75.4611	-7.7878	0.9900	70.2641	10.6679	0.9848
	a_1	346.1070	2.5603	0.0116	348.2524	6.5464	0.0046	348.1475	6.2619	-0.0030
	a_2	0.2192	0.7572		0.0494	0.1433		-0.3007	-0.3606	
19	a_0	82.9249	-23.1378	0.9781	63.7597	-1.1018	0.9946	58.1096	16.5656	0.9818
	a_1	346.5604	4.0813	0.0113	348.3512	6.8347	0.0025	347.5346	5.5403	-0.0045
	a_2	0.2762	0.6384		-0.1039	-0.0354		-0.3697	-0.5367	
20	a_0	69.7608	-18.4214	0.9895	52.0040	5.6935	0.9971	45.2751	21.5658	0.9774
	a_1	347.1247	5.3622	0.0101	348.1399	6.7658	-0.0002	346.7780	4.4625	-0.0060
	a_2	0.2395	0.4821		-0.2466	-0.2283		-0.3396	-0.6992	
21	a_0	57.1229	-12.5808	0.9996	39.8952	12.2268	0.9969	31.7166	25.3267	0.9713
	a_1	347.6108	6.3294	0.0077	347.6379	6.3098	-0.0033	346.0815	3.0541	-0.0076
	a_2	0.1374	0.3006		-0.3488	-0.4323		-0.1871	-0.8144	
22	a_0	44.8682	-5.9550	1.0074	27.1839	18.1002	0.9936	17.6160	27.5657	0.9637
	a_1	347.8881	6.9330	0.0043	346.9250	5.4422	-0.0064	345.6999	1.4115	-0.0089
	a_2	0.0015	0.0989		-0.3692	-0.6304		0.0564	-0.8471	
23	a_0	32.7549	1.0724	1.0117	13.7418	22.9087	0.9872	3.3773	28.1313	0.9548
	a_1	347.8898	7.1325	0.0001	346.1676	4.1724	-0.0093	345.8203	-0.2954	-0.0097
	a_2	-0.1400	-0.1189		-0.2693	-0.7896		0.3014	-0.7860	
24	a_0	20.5022	8.0814	1.0118	359.6446	26.2899	0.9779	349.5024	27.0526	0.9450
	a_1	347.6038	6.8945	-0.0045	345.6154	2.5791	-0.0114	346.4401	-1.8744	-0.0100
	a_2	-0.2581	-0.3443		-0.0514	-0.8696		0.4594	-0.6551	
25	a_0	7.8467	14.6271	1.0072	345.2140	27.9999	0.9664	336.4027	24.5264	0.9350
	a_1	347.0759	6.2024	-0.0088	345.5137	0.8250	-0.0126	347.3751	-3.1853	-0.0096
	a_2	-0.3165	-0.5599		0.2135	-0.8478		0.5006	-0.4967	
26	a_0	354.6070	20.2659	0.9984	330.9456	27.9794	0.9538	324.2775	20.8474	0.9254
	a_1	346.4263	5.0745	-0.0122	345.9557	-0.8804	-0.0127	348.3867	-4.1757	-0.0085
	a_2	-0.2765	-0.7383		0.4225	-0.7386		0.4491	-0.3459	
27	a_0	340.7600	24.5999	0.9861	317.3257	26.3636	0.9411	313.1112	16.3284	0.9168
	a_1	345.8578	3.5854	-0.0143	346.8189	-2.3603	-0.0118	349.2899	-4.8634	-0.0068
	a_2	-0.1220	-0.8468		0.5142	-0.5850		0.3450	-0.2194	
28	a_0	326.5006	27.3381	0.9717	304.6585	23.4216	0.9292	302.7435	11.2478	0.9100
	a_1	345.6087	1.8777	-0.0150	347.8608	-3.5279	-0.0102	349.9817	-5.2986	-0.0045
	a_2	0.1089	-0.8624		0.4951	-0.4296		0.2182	-0.1185	
29	a_0	312.2227	28.3549	0.9566	293.0127	19.4668	0.9190	292.9405	5.8324	0.9055
	a_1	345.8350	0.1419	-0.0145	348.8581	-4.3829	-0.0079	350.4183	-5.5334	-0.0019
	a_2	0.3298	-0.7894		0.4053	-0.2962		0.0838	-0.0364	
30	a_0	298.3904	27.7101	0.9421	282.2735	14.7900	0.9112	283.4397	0.2642	0.9036
	a_1	346.5108	-1.4418	-0.0130	349.6714	-4.9714	-0.0052	350.5859	-5.6056	0.0010
	a_2	0.4639	-0.6600		0.2824	-0.1897		-0.0529	0.0373	
31	a_0	285.3657	25.6113	0.9290				273.9698	-5.3025	0.9046
	a_1	347.4532	-2.7614	-0.0108				350.4798	-5.5322	0.0039
	a_2	0.4901	-0.5143					-0.1905	0.1142	

		January			February			March		
Day		GHA °	DEC °	HP °	GHA °	DEC °	HP °	GHA °	DEC °	HP °
1	a_0	264.2563	-10.7186	0.9085	250.3028	-26.0977	0.9360	268.7814	-27.4676	0.9333
	a_1	350.0978	-5.3066	0.0068	347.1013	-2.5198	0.0140	346.8560	-1.5954	0.0139
	a_2	-0.3253	0.2060		-0.4588	0.7013		-0.3313	0.7450	
2	a_0	254.0263	-15.8168	0.9153	236.9488	-27.9136	0.9500	255.3099	-28.3164	0.9472
	a_1	349.4433	-4.8987	0.0093	346.1681	-1.1100	0.0151	346.1855	-0.0970	0.0155
	a_2	-0.4454	0.3226		-0.2929	0.8266		-0.1510	0.8236	
3	a_0	243.0226	-20.3898	0.9246	222.8290	-28.1959	0.9652	241.3486	-27.5896	0.9628
	a_1	348.5435	-4.2573	0.0113	345.5771	0.5564	0.0149	345.8863	1.5618	0.0162
	a_2	-0.5235	0.4681		-0.0493	0.8819		0.0499	0.8327	
4	a_0	231.0428	-24.1757	0.9359	208.3616	-26.7584	0.9802	227.2878	-25.1965	0.9790
	a_1	347.4807	-3.3220	0.0125	345.4881	2.3352	0.0134	345.9983	3.2387	0.0155
	a_2	-0.5167	0.6327		0.1812	0.8385		0.1982	0.7612	
5	a_0	218.0095	-26.8619	0.9485	194.0337	-23.5876	0.9937	213.4854	-21.1996	0.9946
	a_1	346.4283	-2.0513	0.0128	345.8684	4.0238	0.0106	346.4096	4.7701	0.0133
	a_2	-0.3869	0.7852		0.3151	0.7014		0.2511	0.6179	
6	a_0	204.0559	-28.1261	0.9614	180.2175	-18.8665	1.0043	200.1453	-15.8157	1.0080
	a_1	345.6426	-0.4683	0.0121	346.5150	5.4326	0.0066	346.9234	6.0117	0.0096
	a_2	-0.1437	0.8773		0.3288	0.5015		0.2117	0.4223	
7	a_0	189.5605	-27.7173	0.9736	167.0597	-12.9370	1.0109	187.2784	-9.3864	1.0176
	a_1	345.3596	1.3022	0.0103	347.1828	6.4370	0.0020	347.3533	6.8591	0.0046
	a_2	0.1301	0.8690		0.2502	0.2751		0.1095	0.1953	
8	a_0	175.0542	-25.5484	0.9840	154.4900	-6.2297	1.0128	174.7384	-2.3370	1.0222
	a_1	345.6373	3.0533	0.0077	347.6874	6.9854	-0.0026	347.5739	7.2498	-0.0010
	a_2	0.3268	0.7578		0.1216	0.0490		-0.0210	-0.0443	
9	a_0	161.0198	-21.7409	0.9917	142.2961	0.8004	1.0101	162.2887	4.8637	1.0212
	a_1	346.3105	4.5755	0.0046	347.9305	7.0804	-0.0065	347.5290	7.1591	-0.0062
	a_2	0.3924	0.5772		-0.0215	-0.1618		-0.1479	-0.2792	
10	a_0	147.7218	-16.5925	0.9963	130.2023	7.7150	1.0036	149.6679	11.7392	1.0149
	a_1	347.1090	5.7308	0.0014	347.8834	6.7533	-0.0094	347.2252	6.5964	-0.0105
	a_2	0.3426	0.3705		-0.1514	-0.3518		-0.2386	-0.4936	
11	a_0	135.1709	-10.4955	0.9976	117.9325	14.1130	0.9941	136.6541	17.8382	1.0044
	a_1	347.8010	6.4696	-0.0015	347.5724	6.0460	-0.0111	346.7351	5.6023	-0.0133
	a_2	0.2225	0.1675		-0.2405	-0.5194		-0.2580	-0.6717	
12	a_0	123.1913	-3.8623	0.9961	105.2640	19.6367	0.9830	123.1329	22.7662	0.9910
	a_1	348.2480	6.8018	-0.0039	347.0789	5.0025	-0.0117	346.2042	4.2493	-0.0147
	a_2	0.0722	-0.0195		-0.2588	-0.6600		-0.1793	-0.7950	
13	a_0	111.5083	2.9165	0.9921	92.0856	23.9771	0.9712	109.1613	26.2195	0.9762
	a_1	348.3910	6.7608	-0.0056	346.5470	3.6754	-0.0116	345.8353	2.6477	-0.0147
	a_2	-0.0809	-0.1912		-0.1836	-0.7615		-0.0090	-0.8460	
14	a_0	99.8156	9.4827	0.9865	78.4524	26.8901	0.9596	94.9918	28.0218	0.9614
	a_1	348.2244	6.3772	-0.0067	346.1700	2.1429	-0.0109	345.8177	0.9447	-0.0138
	a_2	-0.2148	-0.3532		-0.0219	-0.8062		0.1986	-0.8178	
15	a_0	87.8233	15.5034	0.9798	64.6046	28.2274	0.9488	81.0116	28.1506	0.9476
	a_1	347.7856	5.6695	-0.0073	346.1264	0.5206	-0.0099	346.2255	-0.6985	-0.0123
	a_2	-0.3031	-0.5089		0.1763	-0.7828		0.3680	-0.7244	

GHA°, DEC° = $(a_2x + a_1)x + a_0$ HP° = $a_1x + a_0$ where $x = GMT^h/24$

Day		January GHA °	January DEC °	January HP °	February GHA °	February DEC °	February HP °	March GHA °	March DEC °	March HP °
16	a_0	75.3057	20.6610	0.9724	50.9106	27.9669	0.9389	67.6069	26.7303	0.9353
	a_1	347.1656	4.6485	-0.0076	346.4887	-1.0526	-0.0088	346.9755	-2.1502	-0.0105
	a_2	-0.3125	-0.6519		0.3405	-0.6977		0.4508	-0.5946	
17	a_0	62.1608	24.6554	0.9648	37.7416	26.2191	0.9301	55.0331	23.9884	0.9248
	a_1	346.5247	3.3379	-0.0078	347.1829	-2.4516	-0.0077	347.8883	-3.3386	-0.0085
	a_2	-0.2171	-0.7625		0.4242	-0.5737		0.4458	-0.4570	
18	a_0	48.4723	27.2297	0.9570	25.3488	23.1967	0.9224	43.3658	20.1955	0.9163
	a_1	346.0801	1.8022	-0.0079	348.0423	-3.5990	-0.0066	348.7863	-4.2503	-0.0066
	a_2	-0.0267	-0.8134		0.4237	-0.4369		0.3796	-0.3291	
19	a_0	34.5303	28.2191	0.9490	13.8135	19.1635	0.9158	32.5298	15.6185	0.9097
	a_1	346.0283	0.1640	-0.0080	348.8963	-4.4710	-0.0056	349.5484	-4.9060	-0.0049
	a_2	0.1976	-0.7864		0.3625	-0.3061		0.2814	-0.2165	
20	a_0	20.7598	27.5987	0.9410	3.0702	14.3888	0.9102	22.3572	10.4980	0.9048
	a_1	346.4356	-1.4173	-0.0080	349.6243	-5.0807	-0.0045	350.1119	-5.3373	-0.0033
	a_2	0.3740	-0.6888		0.2676	-0.1894		0.1706	-0.1176	
21	a_0	7.5711	25.4955	0.9330	352.9598	9.1209	0.9057	12.6374	5.0450	0.9015
	a_1	347.1987	-2.7983	-0.0077	350.1606	-5.4573	-0.0033	350.4530	-5.5716	-0.0018
	a_2	0.4544	-0.5496		0.1579	-0.0876		0.0572	-0.0276	
22	a_0	355.2240	22.1506	0.9253	343.2759	3.5780	0.9024	3.1452	-0.5523	0.8997
	a_1	348.1191	-3.8967	-0.0072	350.4767	-5.6310	-0.0019	350.5669	-5.6265	-0.0004
	a_2	0.4417	-0.4013		0.0430	0.0028		-0.0551	0.0593	
23	a_0	343.7833	17.8555	0.9180	333.7931	-2.0484	0.9005	353.6546	-6.1177	0.8993
	a_1	349.0091	-4.6963	-0.0064	350.5626	-5.6248	-0.0002	350.4558	-5.5085	0.0011
	a_2	0.3662	-0.2654		-0.0733	0.0872		-0.1639	0.1482	
24	a_0	333.1564	12.8962	0.9116	324.2799	-7.5842	0.9003	343.9443	-11.4760	0.9004
	a_1	349.7444	-5.2238	-0.0051	350.4154	-5.4508	0.0018	350.1261	-5.2132	0.0026
	a_2	0.2583	-0.1504		-0.1889	0.1724		-0.2650	0.2440	
25	a_0	323.1566	7.5241	0.9065	314.5041	-12.8607	0.9021	333.8037	-16.4429	0.9030
	a_1	350.2619	-5.5219	-0.0033	350.0361	-5.1075	0.0040	349.5924	-4.7265	0.0043
	a_2	0.1368	-0.0549		-0.2990	0.2653		-0.3484	0.3498	
26	a_0	313.5527	1.9490	0.9032	304.2392	-17.7006	0.9061	323.0467	-20.8172	0.9073
	a_1	350.5359	-5.6303	-0.0012	349.4343	-4.5791	0.0065	348.8887	-4.0279	0.0062
	a_2	0.0105	0.0271		-0.3921	0.3719		-0.3969	0.4653	
27	a_0	304.0964	-3.6525	0.9020	293.2803	-21.9053	0.9126	311.5387	-24.3773	0.9135
	a_1	350.5569	-5.5760	0.0013	348.6424	-3.8373	0.0091	348.0850	-3.0966	0.0082
	a_2	-0.1180	0.1046		-0.4468	0.4940		-0.3874	0.5839	
28	a_0	294.5326	-9.1222	0.9033	281.4762	-25.2458	0.9217	299.2380	-26.8878	0.9218
	a_1	350.3205	-5.3683	0.0040	347.7369	-2.8492	0.0116	347.2991	-1.9254	0.0103
	a_2	-0.2465	0.1873		-0.4338	0.6250		-0.3038	0.6901	
29	a_0	284.6040	-14.3012	0.9073				286.2364	-28.1216	0.9321
	a_1	349.8254	-4.9965	0.0069				346.6846	-0.5385	0.0124
	a_2	-0.3685	0.2852					-0.1550	0.7625	
30	a_0	274.0588	-19.0099	0.9141				272.7695	-27.8972	0.9445
	a_1	349.0829	-4.4294	0.0096				346.3761	0.9953	0.0141
	a_2	-0.4669	0.4060					0.0154	0.7818	
31	a_0	262.6739	-23.0303	0.9238				259.1637	-26.1210	0.9586
	a_1	348.1380	-3.6200	0.0121				346.4164	2.5676	0.0151
	a_2	-0.5102	0.5496					0.1484	0.7405	

Table 2: MOON, 1989

Day		April GHA °	April DEC °	April HP °	May GHA °	May DEC °	May HP °	June GHA °	June DEC °	June HP °
1	a_0	245.7297	-22.8148	0.9738	239.1944	-7.9970	0.9949	221.3863	16.2907	1.0031
	a_1	346.7259	4.0559	0.0152	348.0810	6.5446	0.0104	346.9294	5.7518	-0.0010
	a_2	0.2038	0.6437		-0.0210	0.1924		-0.4481	-0.5682	
2	a_0	232.6588	-18.1182	0.9891	227.2513	-1.2640	1.0053	207.8690	21.4701	1.0021
	a_1	347.1445	5.3491	0.0140	348.0396	6.9335	0.0074	346.0118	4.6090	-0.0047
	a_2	0.1765	0.5012		-0.1685	-0.0027		-0.3837	-0.7719	
3	a_0	219.9778	-12.2716	1.0032	215.1195	5.6622	1.0127	193.5015	25.3045	0.9974
	a_1	347.5046	6.3564	0.0113	347.6988	6.9316	0.0033	345.2242	3.0508	-0.0081
	a_2	0.0857	0.3209		-0.3074	-0.2264		-0.1695	-0.9056	
4	a_0	207.5655	-5.5987	1.0146	202.5090	12.3625	1.0160	178.5625	27.4493	0.9892
	a_1	347.6790	7.0023	0.0070	347.0735	6.4796	-0.0015	344.8803	1.2210	-0.0110
	a_2	-0.0415	0.1083		-0.3996	-0.4680		0.1383	-0.9236	
5	a_0	195.2001	1.5070	1.0216	189.1830	18.3693	1.0144	163.5868	27.7488	0.9782
	a_1	347.5948	7.2213	0.0018	346.2560	5.5386	-0.0064	345.1713	-0.6409	-0.0128
	a_2	-0.1769	-0.1289		-0.3932	-0.6985		0.4142	-0.8235	
6	a_0	182.6157	8.5943	1.0234	175.0488	23.2060	1.0080	149.1754	26.2879	0.9653
	a_1	347.2345	6.9632	-0.0038	345.4483	4.1292	-0.0105	346.0220	-2.2938	-0.0135
	a_2	-0.2874	-0.3759		-0.2460	-0.8708		0.5597	-0.6506	
7	a_0	169.5619	15.1768	1.0195	160.2566	26.4630	0.9974	135.7573	23.3473	0.9517
	a_1	346.6466	6.2068	-0.0089	344.9444	2.3696	-0.0136	347.1591	-3.5934	-0.0132
	a_2	-0.3312	-0.6082		0.0236	-0.9371		0.5654	-0.4632	
8	a_0	155.8786	20.7715	1.0105	145.2306	27.8966	0.9838	123.4801	19.2941	0.9386
	a_1	345.9662	4.9803	-0.0129	344.9982	0.4788	-0.0152	348.2992	-4.5149	-0.0118
	a_2	-0.2661	-0.7924		0.3151	-0.8816		0.4785	-0.2989	
9	a_0	141.5825	24.9574	0.9975	130.5480	27.4968	0.9686	112.2551	14.4831	0.9267
	a_1	345.4188	3.3810	-0.0152	345.6482	-1.2932	-0.0154	349.2593	-5.1075	-0.0098
	a_2	-0.0816	-0.8927		0.5129	-0.7362		0.3491	-0.1691	
10	a_0	126.9248	27.4457	0.9822	116.7104	25.4710	0.9532	101.8605	9.2086	0.9169
	a_1	345.2529	1.5803	-0.0161	346.6937	-2.7664	-0.0144	349.9576	-5.4419	-0.0073
	a_2	0.1693	-0.8891		0.5714	-0.5575		0.2079	-0.0692	
11	a_0	112.3515	28.1389	0.9661	103.9745	22.1506	0.9387	92.0231	3.6992	0.9096
	a_1	345.6032	-0.2086	-0.0155	347.8488	-3.8774	-0.0126	350.3726	-5.5784	-0.0045
	a_2	0.3894	-0.7937		0.5190	-0.3900		0.0680	0.0117	
12	a_0	98.3464	27.1396	0.9505	92.3400	17.8860	0.9262	82.4609	-1.8660	0.9051
	a_1	346.3994	-1.8001	-0.0141	348.8921	-4.6526	-0.0102	350.5076	-5.5548	-0.0017
	a_2	0.5060	-0.6464		0.4068	-0.2528		-0.0668	0.0859	
13	a_0	85.2520	24.6963	0.9364	81.6361	12.9829	0.9160	72.8990	-7.3333	0.9034
	a_1	347.4257	-3.0917	-0.0120	349.7068	-5.1541	-0.0075	350.3726	-5.3844	0.0011
	a_2	0.5112	-0.4895		0.2728	-0.1453		-0.1945	0.1646	
14	a_0	73.1875	21.1181	0.9244	71.6128	7.6854	0.9085	63.0748	-12.5512	0.9045
	a_1	348.4562	-4.0672	-0.0096	350.2519	-5.4423	-0.0048	349.9809	-5.0577	0.0036
	a_2	0.4393	-0.3488		0.1364	-0.0585		-0.3097	0.2571	
15	a_0	62.0807	16.7045	0.9149	61.9984	2.1862	0.9037	52.7442	-17.3495	0.9081
	a_1	349.3380	-4.7612	-0.0071	350.5237	-5.5586	-0.0022	349.3560	-4.5463	0.0057
	a_2	0.3291	-0.2316		0.0045	0.0188		-0.3982	0.3687	

GHA°, DEC° = $(a_2x + a_1)x + a_0$ HP° = $a_1x + a_0$ where $x = GMT^h/24$

Day		April GHA °	April DEC °	April HP °	May GHA °	May DEC °	May HP °	June GHA °	June DEC °	June HP °
16	a_0	51.7452	11.7138	0.9077	52.5240	-3.3520	0.9016	41.7012	-21.5245	0.9138
	a_1	349.9967	-5.2219	-0.0048	350.5315	-5.5217	0.0002	348.5497	-3.8105	0.0074
	a_2	0.2063	-0.1342		-0.1210	0.0972		-0.4351	0.4971	
17	a_0	41.9456	6.3594	0.9029	42.9321	-8.7747	0.9018	29.8167	-24.8352	0.9212
	a_1	350.4087	-5.4892	-0.0026	350.2877	-5.3292	0.0023	347.6659	-2.8147	0.0085
	a_2	0.0829	-0.0488		-0.2372	0.1853		-0.3906	0.6279	
18	a_0	32.4346	0.8231	0.9003	32.9805	-13.9164	0.9042	17.0949	-27.0198	0.9297
	a_1	350.5735	-5.5867	-0.0007	349.8096	-4.9610	0.0041	346.8718	-1.5526	0.0091
	a_2	-0.0366	0.0333		-0.3361	0.2890		-0.2514	0.7330	
19	a_0	22.9691	-4.7285	0.8997	22.4527	-18.5859	0.9083	3.7195	-27.8384	0.9388
	a_1	350.4991	-5.5211	0.0010	349.1306	-4.3849	0.0055	346.3641	-0.0762	0.0091
	a_2	-0.1504	0.1192		-0.4004	0.4090		-0.0468	0.7808	
20	a_0	13.3157	-10.1284	0.9007	11.1828	-22.5591	0.9137	350.0409	-27.1344	0.9480
	a_1	350.1963	-5.2842	0.0025	348.3191	-3.5669	0.0065	346.2772	1.4965	0.0087
	a_2	-0.2544	0.2148		-0.4050	0.5376		0.1530	0.7535	
21	a_0	3.2558	-15.1956	0.9032	359.0985	-25.5859	0.9203	336.4737	-24.8864	0.9567
	a_1	349.6836	-4.8563	0.0038	347.4964	-2.4881	0.0073	346.5974	3.0116	0.0080
	a_2	-0.3392	0.3229		-0.3272	0.6560		0.2790	0.6586	
22	a_0	352.5992	-19.7265	0.9071	346.2710	-27.4163	0.9276	323.3507	-21.2190	0.9648
	a_1	348.9981	-4.2115	0.0051	346.8327	-1.1686	0.0079	347.1699	4.3328	0.0071
	a_2	-0.3876	0.4418		-0.1698	0.7373		0.3044	0.5211	
23	a_0	341.2099	-23.4938	0.9122	332.9379	-27.8472	0.9355	310.8238	-16.3682	0.9719
	a_1	348.2127	-3.3271	0.0064	346.4929	0.3158	0.0084	347.7890	5.3760	0.0061
	a_2	-0.3770	0.5627		0.0240	0.7585		0.2438	0.3661	
24	a_0	329.0473	-26.2560	0.9186	319.4581	-26.7738	0.9439	298.8542	-10.6294	0.9779
	a_1	347.4476	-2.1978	0.0077	346.5502	1.8417	0.0087	348.2825	6.1080	0.0050
	a_2	-0.2914	0.6684		0.1865	0.7143		0.1273	0.2078	
25	a_0	316.2066	-27.7840	0.9263	306.1965	-24.2197	0.9527	287.2610	-4.3170	0.9829
	a_1	346.8578	-0.8539	0.0091	346.9371	3.2761	0.0090	348.5395	6.5237	0.0039
	a_2	-0.1405	0.7371		0.2675	0.6187		-0.0185	0.0485	
26	a_0	302.9274	-27.9005	0.9353	293.4010	-20.3275	0.9617	275.7787	2.2518	0.9868
	a_1	346.5783	0.6287	0.0104	347.4846	4.5163	0.0091	348.5021	6.6219	0.0026
	a_2	0.0317	0.7517		0.2571	0.4935		-0.1731	-0.1166	
27	a_0	289.5401	-26.5209	0.9457	281.1411	-15.3207	0.9708	264.1048	8.7534	0.9894
	a_1	346.6510	2.1399	0.0116	348.0073	5.5044	0.0090	348.1519	6.3906	0.0011
	a_2	0.1663	0.7099		0.1746	0.3543		-0.3146	-0.2943	
28	a_0	276.3586	-23.6729	0.9574	269.3204	-9.4651	0.9799	251.9402	14.8457	0.9906
	a_1	346.9961	3.5651	0.0125	348.3612	6.2144	0.0084	347.5128	5.8031	-0.0006
	a_2	0.2236	0.6222		0.0465	0.2060		-0.4102	-0.4851	
29	a_0	263.5779	-19.4881	0.9699	257.7249	-3.0481	0.9883	239.0428	20.1597	0.9900
	a_1	347.4544	4.8132	0.0127	348.4559	6.6286	0.0072	346.6752	4.8303	-0.0026
	a_2	0.1985	0.5028		-0.1049	0.0440		-0.4127	-0.6746	
30	a_0	251.2290	-14.1750	0.9827	246.0727	3.6206	0.9955	225.3080	24.3123	0.9874
	a_1	347.8590	5.8220	0.0121	348.2447	6.7198	0.0052	345.8291	3.4722	-0.0048
	a_2	0.1090	0.3596		-0.2590	-0.1391		-0.2812	-0.8275	
31	a_0				234.0557	10.1971	1.0007			
	a_1				347.7200	6.4447	0.0024			
	a_2				-0.3882	-0.3465				

Table 2: MOON, 1989

Day		July			August			September		
		GHA °	DEC °	HP °	GHA °	DEC °	HP °	GHA °	DEC °	HP °
1	a_0	210.8611	26.9556	0.9826	186.1362	22.1813	0.9440	172.7828	3.3502	0.9111
	a_1	345.2538	1.8021	-0.0070	347.5607	-3.9368	-0.0089	350.2384	-5.6767	-0.0050
	a_2	-0.0269	-0.8975		0.4830	-0.4312		0.0735	0.0252	
2	a_0	196.0939	27.8610	0.9756	174.1784	17.8165	0.9350	163.0923	-2.2995	0.9061
	a_1	345.2044	-0.0090	-0.0089	348.5344	-4.7961	-0.0087	350.3846	-5.6252	-0.0036
	a_2	0.2606	-0.8576		0.4094	-0.2745		-0.0406	0.1136	
3	a_0	181.5632	26.9972	0.9667	163.1198	12.7486	0.9263	153.4340	-7.8093	0.9025
	a_1	345.7437	-1.7346	-0.0103	349.3562	-5.3413	-0.0080	350.3022	-5.3978	-0.0018
	a_2	0.4681	-0.7242		0.2985	-0.1417		-0.1492	0.1985	
4	a_0	167.7767	24.5421	0.9563	152.7719	7.2679	0.9182	143.5849	-13.0067	0.9007
	a_1	346.6994	-3.1854	-0.0111	349.9538	-5.6213	-0.0068	350.0016	-5.0014	0.0003
	a_2	0.5424	-0.5454		0.1743	-0.0330		-0.2475	0.2858	
5	a_0	155.0178	20.8151	0.9452	142.8974	1.6155	0.9114	133.3373	-17.7203	0.9011
	a_1	347.7973	-4.2732	-0.0110	350.3021	-5.6851	-0.0051	349.5024	-4.4308	0.0028
	a_2	0.5037	-0.3677		0.0484	0.0575		-0.3258	0.3802	
6	a_0	143.3167	16.1773	0.9342	133.2453	-4.0104	0.9063	122.5130	-21.7688	0.9038
	a_1	348.8108	-5.0038	-0.0101	350.3982	-5.5693	-0.0029	348.8439	-3.6713	0.0054
	a_2	0.3998	-0.2164		-0.0750	0.1382		-0.3676	0.4825	
7	a_0	132.5246	10.9596	0.9240	123.5660	-9.4398	0.9034	110.9896	-24.9554	0.9092
	a_1	349.6122	-5.4320	-0.0085	350.2469	-5.2933	-0.0004	348.0992	-2.7058	0.0081
	a_2	0.2692	-0.0966		-0.1931	0.2183		-0.3530	0.5873	
8	a_0	122.4032	5.4330	0.9154	113.6176	-14.5129	0.9030	98.7376	-27.0720	0.9173
	a_1	350.1504	-5.6221	-0.0064	349.8582	-4.8582	0.0024	347.3834	-1.5280	0.0108
	a_2	0.1325	-0.0023		-0.2997	0.3063		-0.2700	0.6806	
9	a_0	112.6833	-0.1898	0.9090	103.1744	-19.0627	0.9054	85.8538	-27.9182	0.9281
	a_1	350.4148	-5.6252	-0.0039	349.2537	-4.2477	0.0052	346.8377	-0.1607	0.0131
	a_2	-0.0022	0.0768		-0.3812	0.4079		-0.1306	0.7424	
10	a_0	103.0932	-5.7366	0.9052	92.0462	-22.9001	0.9107	72.5640	-27.3362	0.9413
	a_1	350.4094	-5.4717	-0.0011	348.4823	-3.4333	0.0080	346.5784	1.3323	0.0149
	a_2	-0.1325	0.1522		-0.4151	0.5229		0.0235	0.7543	
11	a_0	93.3675	-11.0544	0.9041	80.1142	-25.8080	0.9187	59.1684	-25.2506	0.9563
	a_1	350.1426	-5.1689	0.0018	347.6398	-2.3863	0.0105	346.6344	2.8493	0.0157
	a_2	-0.2550	0.2345		-0.3754	0.6410		0.1400	0.7079	
12	a_0	83.2528	-15.9867	0.9059	67.3811	-27.5513	0.9292	45.9438	-21.6955	0.9721
	a_1	349.6287	-4.7024	0.0046	346.8775	-1.0992	0.0125	346.9260	4.2725	0.0154
	a_2	-0.3602	0.3322		-0.2514	0.7393		0.1848	0.6061	
13	a_0	72.5199	-20.3545	0.9105	54.0110	-27.9102	0.9417	33.0540	-16.8201	0.9875
	a_1	348.9008	-4.0405	0.0071	346.3702	0.3885	0.0137	347.3057	5.4905	0.0135
	a_2	-0.4283	0.4486		-0.0690	0.7891		0.1542	0.4576	
14	a_0	60.9924	-23.9437	0.9177	40.3159	-26.7330	0.9555	20.5120	-10.8758	1.0011
	a_1	348.0322	-3.1439	0.0092	346.2381	1.9775	0.0139	347.6206	6.4102	0.0102
	a_2	-0.4300	0.5777		0.1104	0.7693		0.0657	0.2717	
15	a_0	48.5965	-26.5074	0.9269	26.6669	-23.9881	0.9694	8.1958	-4.1984	1.0114
	a_1	347.1579	-1.9850	0.0107	346.4717	3.5255	0.0129	347.7548	6.9566	0.0058
	a_2	-0.3396	0.6991		0.2251	0.6780		-0.0560	0.0578	

$GHA^\circ, DEC^\circ = (a_2x + a_1)x + a_0$ $\quad HP^\circ = a_1x + a_0 \quad$ where $x = GMT^h/24$

		July			August			September		
Day		GHA °	DEC °	HP °	GHA °	DEC °	HP °	GHA °	DEC °	HP °
16	a_0	35.4184	-27.7916	0.9377	13.3642	-19.7877	0.9824	355.8919	2.8112	1.0171
	a_1	346.4687	-0.5783	0.0114	346.9352	4.8880	0.0108	347.6415	7.0734	0.0007
	a_2	-0.1618	0.7808		0.2485	0.5302		-0.1842	-0.1729	
17	a_0	21.7297	-27.5888	0.9491	0.5468	-14.3733	0.9933	343.3471	9.7069	1.0178
	a_1	346.1459	0.9948	0.0113	347.4420	5.9518	0.0077	347.2667	6.7262	-0.0042
	a_2	0.0507	0.7928		0.1918	0.3466		-0.2875	-0.4053	
18	a_0	7.9298	-25.8026	0.9605	348.1783	-8.0790	1.0010	330.3254	16.0234	1.0136
	a_1	346.2587	2.5913	0.0104	347.8309	6.6463	0.0040	346.6791	5.9108	-0.0083
	a_2	0.2210	0.7259		0.0828	0.1456		-0.3263	-0.6189	
19	a_0	354.4110	-22.4880	0.9709	336.0892	-1.2914	1.0049	316.6796	21.3119	1.0052
	a_1	346.7159	4.0503	0.0087	347.9978	6.9372	0.0001	346.0096	4.6639	-0.0113
	a_2	0.2975	0.5955		-0.0503	-0.0603		-0.2617	-0.7864	
20	a_0	341.4241	-17.8457	0.9796	324.0340	5.5813	1.0050	302.4311	25.1876	0.9938
	a_1	347.3236	5.2444	0.0065	347.8949	6.8154	-0.0034	345.4720	3.0780	-0.0129
	a_2	0.2767	0.4285		-0.1815	-0.2629		-0.0849	-0.8780	
21	a_0	329.0226	-12.1766	0.9862	311.7453	12.1298	1.0015	287.8232	27.3876	0.9809
	a_1	347.8849	6.1018	0.0041	347.5250	6.2874	-0.0062	345.2997	1.3080	-0.0133
	a_2	0.1850	0.2484		-0.2824	-0.4560		0.1544	-0.8750	
22	a_0	317.0899	-5.8302	0.9903	298.9871	17.9575	0.9953	273.2816	27.8224	0.9675
	a_1	348.2586	6.5978	0.0017	346.9478	5.3713	-0.0081	345.6193	-0.4524	-0.0129
	a_2	0.0536	0.0681		-0.3177	-0.6302		0.3657	-0.7858	
23	a_0	305.3990	0.8320	0.9920	285.6185	22.6958	0.9871	259.2691	26.5871	0.9546
	a_1	348.3661	6.7334	-0.0004	346.2963	4.1037	-0.0092	346.3672	-2.0286	-0.0118
	a_2	-0.0927	-0.1090		-0.2539	-0.7678		0.4807	-0.6439	
24	a_0	293.6695	7.4527	0.9915	271.6643	26.0302	0.9779	246.1172	23.9178	0.9428
	a_1	348.1777	6.5151	-0.0022	345.7753	2.5573	-0.0097	347.3424	-3.3160	-0.0105
	a_2	-0.2315	-0.2850		-0.0856	-0.8441		0.4900	-0.4872	
25	a_0	281.6136	13.6792	0.9893	257.3587	27.7434	0.9682	233.9483	20.1177	0.9323
	a_1	347.7069	5.9448	-0.0036	345.6019	0.8567	-0.0096	348.3303	-4.2877	-0.0090
	a_2	-0.3355	-0.4612		0.1401	-0.8398		0.4245	-0.3407	
26	a_0	268.9844	19.1593	0.9857	243.1049	27.7620	0.9585	222.7009	15.4920	0.9234
	a_1	347.0222	5.0202	-0.0048	345.8921	-0.8327	-0.0093	349.1825	-4.9660	-0.0075
	a_2	-0.3668	-0.6308		0.3403	-0.7570		0.3207	-0.2133	
27	a_0	255.6415	23.5457	0.9809	229.3396	26.1750	0.9492	212.2017	10.3150	0.9158
	a_1	346.2709	3.7520	-0.0058	346.5880	-2.3518	-0.0089	349.8245	-5.3901	-0.0061
	a_2	-0.2878	-0.7728		0.4515	-0.6218		0.2039	-0.1040	
28	a_0	241.6286	26.5233	0.9751	216.3794	23.2046	0.9404	202.2277	4.8230	0.9097
	a_1	345.6812	2.1949	-0.0067	347.5041	-3.5958	-0.0083	350.2318	-5.5964	-0.0048
	a_2	-0.0938	-0.8538		0.4635	-0.4674		0.0868	-0.0074	
29	a_0	227.2212	27.8646	0.9685	204.3459	19.1445	0.9320	192.5438	-0.7790	0.9050
	a_1	345.4923	0.4735	-0.0075	348.4390	-4.5285	-0.0077	350.4043	-5.6106	-0.0034
	a_2	0.1581	-0.8463		0.4032	-0.3185		-0.0254	0.0823	
30	a_0	212.8760	27.4937	0.9610	193.1861	14.3003	0.9243	182.9205	-6.3054	0.9015
	a_1	345.8207	-1.2301	-0.0082	349.2489	-5.1625	-0.0070	350.3519	-5.4459	-0.0021
	a_2	0.3723	-0.7513		0.3042	-0.1864		-0.1298	0.1706	
31	a_0	199.0714	25.5153	0.9527	182.7368	8.9537	0.9173			
	a_1	346.5827	-2.7378	-0.0087	349.8583	-5.5326	-0.0061			
	a_2	0.4821	-0.5997		0.1902	-0.0730				

Table 2: MOON, 1989

Day		October GHA °	October DEC °	October HP °	November GHA °	November DEC °	November HP °	December GHA °	December DEC °	December HP °
1	a_0	173.1407	-11.5788	0.8994	158.9312	-25.8167	0.9054	150.8339	-26.4895	0.9245
	a_1	350.0899	-5.1052	-0.0005	347.8625	-2.0953	0.0053	347.1304	1.6131	0.0079
	a_2	-0.2221	0.2614		-0.2381	0.6201		0.1548	0.6556	
2	a_0	163.0069	-16.4207	0.8989	146.5582	-27.2908	0.9107	138.1206	-24.2225	0.9324
	a_1	349.6417	-4.5830	0.0012	347.3813	-0.8494	0.0070	347.4512	2.9289	0.0089
	a_2	-0.2938	0.3572		-0.1125	0.6737		0.2261	0.5795	
3	a_0	152.3540	-20.6444	0.9000	133.8298	-27.4664	0.9177	125.7979	-20.7161	0.9414
	a_1	349.0478	-3.8688	0.0031	347.1576	0.5045	0.0088	347.9141	4.0903	0.0099
	a_2	-0.3312	0.4568		0.0274	0.6845		0.2211	0.4790	
4	a_0	141.0709	-24.0545	0.9031	121.0171	-26.2781	0.9265	113.9317	-16.1492	0.9512
	a_1	348.3772	-2.9542	0.0053	347.2199	1.8792	0.0107	348.3643	5.0493	0.0107
	a_2	-0.3183	0.5544		0.1363	0.6519		0.1515	0.3664	
5	a_0	129.1313	-26.4526	0.9084	108.3743	-23.7483	0.9373	102.4451	-10.7361	0.9620
	a_1	347.7322	-1.8423	0.0076	347.5029	3.1872	0.0125	348.6724	5.7833	0.0113
	a_2	-0.2464	0.6385		0.1830	0.5847		0.0361	0.2453	
6	a_0	116.6196	-27.6552	0.9160	96.0598	-19.9783	0.9498	91.1506	-4.7104	0.9733
	a_1	347.2345	-0.5600	0.0101	347.8783	4.3595	0.0139	348.7475	6.2764	0.0113
	a_2	-0.1264	0.6946		0.1606	0.4932		-0.1081	0.1104	
7	a_0	103.7305	-27.5203	0.9261	84.0971	-15.1279	0.9637	79.7866	1.6730	0.9846
	a_1	346.9832	0.8356	0.0124	348.2067	5.3487	0.0147	348.5317	6.5013	0.0105
	a_2	0.0076	0.7104		0.0803	0.3823		-0.2660	-0.0489	
8	a_0	90.7233	-25.9749	0.9385	72.3815	-9.3997	0.9785	68.0493	8.1214	0.9952
	a_1	347.0057	2.2626	0.0144	348.3719	6.1169	0.0145	347.9961	6.4087	0.0087
	a_2	0.1116	0.6818		-0.0414	0.2488		-0.4156	-0.2425	
9	a_0	77.8415	-23.0320	0.9530	60.7090	-3.0374	0.9931	55.6277	14.2829	1.0040
	a_1	347.2389	3.6317	0.0158	348.2912	6.6197	0.0130	347.1536	5.9275	0.0058
	a_2	0.1547	0.6120		-0.1874	0.0840		-0.5165	-0.4703	
10	a_0	65.2345	-18.7904	0.9689	48.8097	3.6621	1.0061	42.2651	19.7352	1.0098
	a_1	347.5577	4.8607	0.0162	347.9148	6.7936	0.0100	346.0991	4.9850	0.0019
	a_2	0.1299	0.5062		-0.3374	-0.1205		-0.5054	-0.7090	
11	a_0	52.9204	-13.4265	0.9852	36.3846	10.3302	1.0162	27.8625	24.0072	1.0118
	a_1	347.8243	5.8780	0.0152	347.2320	6.5571	0.0056	345.0622	3.5551	-0.0025
	a_2	0.0481	0.3659		-0.4570	-0.3639		-0.3253	-0.9034	
12	a_0	40.7903	-7.1862	1.0005	23.1589	16.5180	1.0218	12.6061	26.6572	1.0092
	a_1	347.9245	6.6151	0.0126	346.3003	5.8286	0.0003	344.3974	1.7275	-0.0070
	a_2	-0.0718	0.1893		-0.4904	-0.6246		0.0035	-0.9843	
13	a_0	28.6401	-0.3863	1.0132	8.9713	21.7173	1.0221	357.0142	27.4018	1.0021
	a_1	347.7813	6.9987	0.0085	345.2945	4.5694	-0.0052	344.4144	-0.2616	-0.0108
	a_2	-0.2083	-0.0249		-0.3742	-0.8521		0.3515	-0.9181	
14	a_0	16.2105	6.5825	1.0217	353.8973	25.4320	1.0169	341.7847	26.2257	0.9912
	a_1	347.3600	6.9522	0.0032	344.5266	2.8458	-0.0101	345.1427	-2.1088	-0.0135
	a_2	-0.3324	-0.2702		-0.0955	-0.9789		0.5736	-0.7394	
15	a_0	3.2368	13.2592	1.0249	338.3355	27.2993	1.0067	327.5022	23.3819	0.9776
	a_1	346.6833	6.4108	-0.0025	344.3370	0.8665	-0.0137	346.3131	-3.5881	-0.0149
	a_2	-0.4014	-0.5265		0.2523	-0.9610		0.6245	-0.5209	

GHA°, DEC° $= (a_2x + a_1)x + a_0$ HP° $= a_1x + a_0$ where $x = \mathrm{GMT}^h/24$

Day		October GHA °	October DEC °	October HP °	November GHA °	November DEC °	November HP °	December GHA °	December DEC °	December HP °
16	a_0	349.5195	19.1387	1.0223	322.9304	27.2078	0.9929	314.4384	19.2770	0.9627
	a_1	345.8611	5.3502	-0.0078	344.8629	-1.0693	-0.0158	347.5751	-4.6245	-0.0149
	a_2	-0.3619	-0.7561		0.5197	-0.8172		0.5508	-0.3203	
17	a_0	335.0224	23.7297	1.0144	308.3152	25.3255	0.9770	302.5616	14.3355	0.9477
	a_1	345.1174	3.8230	-0.0120	345.9269	-2.7067	-0.0163	348.6813	-5.2586	-0.0138
	a_2	-0.1819	-0.9093		0.6242	-0.6142		0.4175	-0.1611	
18	a_0	319.9637	26.6426	1.0023	294.8657	22.0088	0.9606	291.6572	8.9184	0.9338
	a_1	344.7461	1.9858	-0.0146	347.1915	-3.9310	-0.0155	349.5165	-5.5754	-0.0118
	a_2	0.1001	-0.9458		0.5869	-0.4154		0.2678	-0.0419	
19	a_0	304.8155	27.6843	0.9876	282.6418	17.6658	0.9451	281.4384	3.3030	0.9220
	a_1	344.9570	0.0793	-0.0157	348.3722	-4.7556	-0.0138	350.0507	-5.6559	-0.0093
	a_2	0.3733	-0.8638		0.4708	-0.2516		0.1214	0.0490	
20	a_0	290.1493	26.9030	0.9718	271.4818	12.6612	0.9313	271.6077	-2.3022	0.9127
	a_1	345.7241	-1.6551	-0.0154	349.3151	-5.2535	-0.0115	350.2918	-5.5566	-0.0065
	a_2	0.5358	-0.7052		0.3270	-0.1263		-0.0153	0.1253	
21	a_0	276.4098	24.5463	0.9563	261.1209	7.2835	0.9198	261.8816	-7.7319	0.9062
	a_1	346.8135	-3.0650	-0.0142	349.9682	-5.5026	-0.0089	350.2591	-5.3063	-0.0036
	a_2	0.5634	-0.5252		0.1819	-0.0297		-0.1398	0.1994	
22	a_0	263.7854	20.9595	0.9421	251.2680	1.7529	0.9109	251.9986	-12.8371	0.9026
	a_1	347.9505	-4.1114	-0.0124	350.3302	-5.5605	-0.0062	349.9762	-4.9089	-0.0008
	a_2	0.4942	-0.3613		0.0451	0.0510		-0.2474	0.2808	
23	a_0	252.2276	16.4896	0.9297	241.6406	-3.7550	0.9047	241.7258	-17.4632	0.9018
	a_1	348.9426	-4.8296	-0.0102	350.4184	-5.4585	-0.0037	349.4760	-4.3491	0.0017
	a_2	0.3773	-0.2262		-0.0805	0.1276		-0.3266	0.3745	
24	a_0	241.5448	11.4361	0.9194	231.9762	-9.0842	0.9011	230.8746	-21.4357	0.9035
	a_1	349.6976	-5.2786	-0.0080	350.2549	-5.2045	-0.0014	348.8144	-3.6011	0.0039
	a_2	0.2462	-0.1172		-0.1921	0.2091		-0.3587	0.4790	
25	a_0	231.4859	6.0421	0.9114	222.0371	-14.0777	0.8997	219.3311	-24.5557	0.9074
	a_1	350.1890	-5.5112	-0.0059	349.8668	-4.7879	0.0006	348.0860	-2.6417	0.0056
	a_2	0.1169	-0.0255		-0.2826	0.3011		-0.3229	0.5833	
26	a_0	221.7892	0.5072	0.9055	211.6201	-18.5623	0.9004	207.0965	-26.6125	0.9130
	a_1	350.4212	-5.5615	-0.0040	349.2950	-4.1869	0.0024	347.4301	-1.4703	0.0069
	a_2	-0.0047	0.0586		-0.3383	0.4043		-0.2118	0.6669	
27	a_0	212.2034	-4.9940	0.9015	200.5768	-22.3427	0.9028	194.3181	-27.4150	0.9199
	a_1	350.4100	-5.4447	-0.0022	348.6092	-3.3780	0.0038	347.0024	-0.1285	0.0077
	a_2	-0.1160	0.1427		-0.3398	0.5116		-0.0483	0.7067	
28	a_0	202.4953	-10.2941	0.8993	188.8475	-25.2071	0.9066	181.2756	-26.8371	0.9277
	a_1	350.1753	-5.1602	-0.0006	347.9192	-2.3515	0.0050	346.9103	1.2935	0.0081
	a_2	-0.2134	0.2325		-0.2723	0.6081		0.1154	0.6887	
29	a_0	192.4557	-15.2198	0.8986	176.4972	-26.9491	0.9115	168.3036	-24.8564	0.9358
	a_1	349.7444	-4.6963	0.0008	347.3673	-1.1291	0.0060	347.1520	2.6777	0.0082
	a_2	-0.2884	0.3301		-0.1410	0.6733		0.2258	0.6164	
30	a_0	181.9108	-19.5839	0.8995	163.7269	-27.4045	0.9175	155.6821	-21.5645	0.9440
	a_1	349.1611	-4.0365	0.0023	347.0852	0.2253	0.0070	347.6155	3.9144	0.0081
	a_2	-0.3272	0.4334		0.0189	0.6904		0.2567	0.5067	
31	a_0	170.7450	-23.1849	0.9017				143.5534	-17.1460	0.9521
	a_1	348.4980	-3.1683	0.0037				348.1384	4.9293	0.0078
	a_2	-0.3134	0.5347					0.2133	0.3786	

Table 2: MOON, 1990

Day		January GHA °	January DEC °	January HP °	February GHA °	February DEC °	February HP °	March GHA °	March DEC °	March HP °
1	a_0	131.9030	-11.8409	0.9599	115.6830	11.5647	0.9865	133.4595	15.7680	0.9987
	a_1	348.5712	5.6867	0.0075	347.9400	6.0362	0.0008	347.0847	5.6505	-0.0033
	a_2	0.1158	0.2439		-0.3451	-0.3769		-0.3636	-0.5541	
2	a_0	120.5872	-5.9133	0.9675	103.2766	17.2203	0.9873	120.1812	20.8608	0.9954
	a_1	348.8062	6.1750	0.0072	347.2386	5.2819	-0.0005	346.3416	4.5364	-0.0054
	a_2	-0.0156	0.1055		-0.4087	-0.5579		-0.3301	-0.7253	
3	a_0	109.3747	0.3642	0.9747	90.1072	21.9408	0.9868	106.1958	24.6695	0.9900
	a_1	348.7763	6.3876	0.0067	346.4039	4.1620	-0.0017	345.6654	3.0754	-0.0068
	a_2	-0.1646	-0.0414		-0.3741	-0.7260		-0.1831	-0.8420	
4	a_0	97.9833	6.7070	0.9814	76.1402	25.3744	0.9852	91.6829	26.9022	0.9832
	a_1	348.4454	6.3077	0.0059	345.6380	2.7002	-0.0029	345.2924	1.3779	-0.0076
	a_2	-0.3145	-0.2050		-0.2145	-0.8481		0.0482	-0.8756	
5	a_0	86.1117	12.8058	0.9873	61.5690	27.2257	0.9823	77.0282	27.4057	0.9756
	a_1	347.8097	5.9008	0.0048	345.2014	0.9896	-0.0042	345.3958	-0.3858	-0.0080
	a_2	-0.4380	-0.3914		0.0388	-0.8862		0.2810	-0.8177	
6	a_0	73.4824	18.3110	0.9921	46.8144	27.3304	0.9780	62.7082	26.2048	0.9675
	a_1	346.9189	5.1189	0.0031	345.2871	-0.7968	-0.0056	345.9736	-2.0288	-0.0082
	a_2	-0.4895	-0.5947		0.2927	-0.8239		0.4331	-0.6894	
7	a_0	59.9134	22.8312	0.9952	32.3976	25.7126	0.9724	49.1158	23.4900	0.9593
	a_1	345.9177	3.9242	0.0007	345.8902	-2.4533	-0.0071	346.8551	-3.4096	-0.0082
	a_2	-0.4125	-0.7865		0.4553	-0.6814		0.4755	-0.5272	
8	a_0	45.4233	25.9663	0.9960	18.7441	22.5817	0.9653	36.4455	19.5566	0.9511
	a_1	345.0726	2.3374	-0.0021	346.8176	-3.8181	-0.0083	347.8162	-4.4624	-0.0082
	a_2	-0.1831	-0.9146		0.4969	-0.5007		0.4303	-0.3627	
9	a_0	30.3193	27.3889	0.9939	6.0576	18.2666	0.9570	24.6900	14.7346	0.9429
	a_1	344.7027	0.4894	-0.0051	347.8218	-4.8172	-0.0093	348.6813	-5.1849	-0.0081
	a_2	0.1359	-0.9284		0.4444	-0.3199		0.3353	-0.2126	
10	a_0	15.1637	26.9521	0.9888	354.3217	13.1328	0.9477	13.7041	9.3399	0.9349
	a_1	344.9903	-1.3830	-0.0080	348.7152	-5.4529	-0.0098	349.3529	-5.6071	-0.0078
	a_2	0.4136	-0.8197		0.3409	-0.1601		0.2205	-0.0816	
11	a_0	0.5706	24.7533	0.9808	343.3752	7.5226	0.9379	3.2751	3.6536	0.9270
	a_1	345.8404	-3.0291	-0.0104	349.3979	-5.7690	-0.0096	349.7935	-5.7678	-0.0073
	a_2	0.5537	-0.6315		0.2182	-0.0272		0.1033	0.0322	
12	a_0	346.9649	21.0970	0.9703	332.9887	1.7287	0.9282	353.1695	-2.0799	0.9197
	a_1	346.9653	-4.2910	-0.0120	349.8337	-5.8202	-0.0088	349.9986	-5.7018	-0.0066
	a_2	0.5525	-0.4238		0.0937	0.0817		-0.0083	0.1330	
13	a_0	334.4809	16.3861	0.9583	322.9136	-4.0078	0.9194	343.1577	-7.6467	0.9131
	a_1	348.0793	-5.1334	-0.0126	350.0198	-5.6547	-0.0074	349.9800	-5.4348	-0.0054
	a_2	0.4610	-0.2378		-0.0247	0.1737		-0.1092	0.2257	
14	a_0	323.0185	11.0180	0.9457	312.9064	-9.4870	0.9120	333.0267	-12.8540	0.9076
	a_1	349.0041	-5.6032	-0.0122	349.9683	-5.3064	-0.0054	349.7585	-4.9829	-0.0039
	a_2	0.3301	-0.0885		-0.1327	0.2569		-0.1939	0.3145	
15	a_0	312.3498	5.3287	0.9335	302.7402	-14.5348	0.9066	322.5900	-17.5205	0.9038
	a_1	349.6641	-5.7757	-0.0109	349.6998	-4.7927	-0.0029	349.3661	-4.3536	-0.0019
	a_2	0.1901	0.0263		-0.2240	0.3386		-0.2536	0.4020	

$GHA^\circ, DEC^\circ = (a_2x + a_1)x + a_0$ $HP^\circ = a_1x + a_0$ where $x = GMT^h/24$

Day		January			February			March		
		GHA °	DEC °	HP °	GHA °	DEC °	HP °	GHA °	DEC °	HP °
16	a_0	302.2012	-0.4188	0.9226	292.2146	-18.9871	0.9037	311.7020	-21.4703	0.9019
	a_1	350.0431	-5.7203	-0.0088	349.2468	-4.1160	-0.0001	348.8526	-3.5489	0.0004
	a_2	0.0543	0.1167		-0.2880	0.4237		-0.2762	0.4876	
17	a_0	292.2959	-6.0207	0.9137	281.1729	-22.6775	0.9036	300.2789	-24.5300	0.9023
	a_1	350.1499	-5.4857	-0.0063	348.6634	-3.2686	0.0028	348.2926	-2.5721	0.0030
	a_2	-0.0724	0.1947		-0.3097	0.5121		-0.2509	0.5662	
18	a_0	282.3711	-11.3101	0.9074	269.5274	-25.4322	0.9064	288.3221	-26.5345	0.9053
	a_1	350.0027	-5.0967	-0.0035	348.0351	-2.2428	0.0057	347.7844	-1.4364	0.0058
	a_2	-0.1860	0.2710		-0.2749	0.5970		-0.1762	0.6287	
19	a_0	272.1858	-16.1341	0.9040	257.2896	-27.0766	0.9121	275.9325	-27.3414	0.9112
	a_1	349.6266	-4.5560	-0.0005	347.4779	-1.0449	0.0084	347.4293	-0.1743	0.0086
	a_2	-0.2789	0.3539		-0.1816	0.6647		-0.0680	0.6641	
20	a_0	261.5323	-20.3342	0.9035	244.5885	-27.4561	0.9205	263.2961	-26.8516	0.9198
	a_1	349.0622	-3.8495	0.0024	347.1120	0.2905	0.0107	347.2960	1.1594	0.0112
	a_2	-0.3362	0.4468		-0.0508	0.6987		0.0416	0.6647	
21	a_0	250.2583	-23.7348	0.9059	231.6524	-26.4672	0.9312	250.6353	-25.0282	0.9311
	a_1	348.3802	-2.9558	0.0051	347.0144	1.6950	0.0123	347.3861	2.4940	0.0134
	a_2	-0.3384	0.5458		0.0768	0.6867		0.1181	0.6282	
22	a_0	238.3015	-26.1430	0.9110	218.7454	-24.0867	0.9436	238.1399	-21.9075	0.9445
	a_1	347.6928	-1.8616	0.0073	347.1768	3.0752	0.0131	347.6308	3.7551	0.0148
	a_2	-0.2701	0.6371		0.1606	0.6262		0.1405	0.5573	
23	a_0	225.7269	-27.3663	0.9183	206.0832	-20.3875	0.9567	225.9106	-17.5972	0.9594
	a_1	347.1451	-0.5815	0.0091	347.5078	4.3331	0.0128	347.9196	4.8741	0.0152
	a_2	-0.1371	0.6996		0.1797	0.5227		0.1062	0.4549	
24	a_0	212.7383	-27.2479	0.9274	193.7698	-15.5345	0.9696	213.9347	-12.2710	0.9747
	a_1	346.8709	0.8259	0.0101	347.8754	5.3827	0.0115	348.1376	5.7883	0.0143
	a_2	0.0241	0.7130		0.1365	0.3852		0.0249	0.3211	
25	a_0	199.6361	-25.7099	0.9376	181.7798	-9.7701	0.9812	202.0949	-6.1650	0.9891
	a_1	346.9267	2.2602	0.0104	348.1538	6.1560	0.0092	348.1907	6.4353	0.0120
	a_2	0.1599	0.6686		0.0465	0.2218		-0.0883	0.1547	
26	a_0	186.7242	-22.7830	0.9481	169.9776	-3.3961	0.9904	190.1946	0.4208	1.0012
	a_1	347.2577	3.6037	0.0100	348.2493	6.6017	0.0061	348.0144	6.7489	0.0084
	a_2	0.2295	0.5732		-0.0713	0.0392		-0.2153	-0.0444	
27	a_0	174.2113	-18.6087	0.9581	158.1529	3.2408	0.9965	177.9913	7.1207	1.0096
	a_1	347.7273	4.7539	0.0090	348.1060	6.6813	0.0028	347.5798	6.6627	0.0040
	a_2	0.2230	0.4426		-0.1971	-0.1564		-0.3313	-0.2700	
28	a_0	162.1602	-13.4153	0.9671	146.0596	9.7615	0.9993	165.2384	13.5086	1.0136
	a_1	348.1808	5.6408	0.0075	347.7070	6.3685	-0.0005	346.9066	6.1221	-0.0007
	a_2	0.1538	0.2927		-0.3059	-0.3579		-0.3987	-0.5052	
29	a_0	150.4926	-7.4851	0.9746				151.7469	19.1211	1.0128
	a_1	348.4928	6.2268	0.0057				346.0919	5.1055	-0.0051
	a_2	0.0430	0.1337					-0.3697	-0.7194	
30	a_0	139.0256	-1.1279	0.9803				137.4723	23.5041	1.0077
	a_1	348.5805	6.4948	0.0039				345.3338	3.6541	-0.0085
	a_2	-0.0900	-0.0302					-0.2125	-0.8704	
31	a_0	127.5132	5.3331	0.9843				122.5989	26.2868	0.9992
	a_1	348.3994	6.4352	0.0023				344.9003	1.8967	-0.0107
	a_2	-0.2271	-0.1999					0.0454	-0.9200	

Day		April			May			June		
		GHA °	DEC °	HP °	GHA °	DEC °	HP °	GHA °	DEC °	HP °
1	a_0	107.5500	27.2648	0.9884	96.3225	21.2065	0.9670	81.5784	1.8447	0.9280
	a_1	344.9990	0.0421	-0.0119	347.1171	-4.0443	-0.0144	349.8776	-5.6571	-0.0104
	a_2	0.3087	-0.8593		0.5586	-0.4107		0.1012	0.0783	
2	a_0	92.8613	26.4504	0.9765	83.9959	16.7550	0.9526	71.5545	-3.7325	0.9176
	a_1	345.6347	-1.6845	-0.0120	348.2408	-4.8601	-0.0130	350.0775	-5.4995	-0.0079
	a_2	0.4794	-0.7187		0.4449	-0.2461		-0.0283	0.1592	
3	a_0	78.9764	24.0508	0.9644	72.6787	11.6514	0.9395	61.6013	-9.0711	0.9096
	a_1	346.6110	-3.1230	-0.0115	349.1317	-5.3477	-0.0113	350.0177	-5.1813	-0.0055
	a_2	0.5242	-0.5459		0.3057	-0.1158		-0.1397	0.2383	
4	a_0	66.1105	20.3853	0.9529	62.1132	6.1902	0.9282	51.4775	-14.0124	0.9041
	a_1	347.6702	-4.2122	-0.0106	349.7419	-5.5761	-0.0093	349.7338	-4.7056	-0.0033
	a_2	0.4713	-0.3783		0.1662	-0.0110		-0.2266	0.3219	
5	a_0	54.2498	15.7979	0.9423	52.0186	0.6050	0.9189	40.9837	-18.3942	0.9009
	a_1	348.6173	-4.9651	-0.0095	350.0723	-5.5966	-0.0074	349.2742	-4.0623	-0.0012
	a_2	0.3659	-0.2321		0.0369	0.0793		-0.2775	0.4114	
6	a_0	43.2304	10.6033	0.9328	42.1254	-4.9105	0.9115	29.9804	-22.0433	0.8996
	a_1	349.3498	-5.4261	-0.0083	350.1437	-5.4377	-0.0056	348.7109	-3.2388	0.0006
	a_2	0.2421	-0.1083		-0.0782	0.1647		-0.2777	0.5012	
7	a_0	32.8197	5.0712	0.9245	32.1888	-10.1818	0.9059	18.4148	-24.7793	0.9002
	a_1	349.8332	-5.6407	-0.0072	349.9840	-5.1088	-0.0038	348.1467	-2.2334	0.0021
	a_2	0.1183	-0.0014		-0.1748	0.2517		-0.2176	0.5791	
8	a_0	22.7687	-0.5689	0.9174	21.9965	-15.0370	0.9021	6.3463	-26.4326	0.9024
	a_1	350.0680	-5.6423	-0.0060	349.6296	-4.6059	-0.0023	347.7056	-1.0700	0.0036
	a_2	0.0025	0.0956		-0.2450	0.3430		-0.1046	0.6291	
9	a_0	12.8371	-6.1137	0.9113	11.3805	-19.2979	0.8998	353.9501	-26.8733	0.9059
	a_1	350.0708	-5.4507	-0.0049	349.1330	-3.9196	-0.0008	347.4963	0.1947	0.0049
	a_2	-0.1008	0.1884		-0.2766	0.4363		0.0308	0.6387	
10	a_0	2.8053	-11.3741	0.9065	0.2372	-22.7794	0.8991	341.4796	-26.0405	0.9108
	a_1	349.8660	-5.0739	-0.0036	348.5717	-3.0452	0.0007	347.5640	1.4780	0.0061
	a_2	-0.1864	0.2807		-0.2571	0.5239		0.1462	0.6055	
11	a_0	352.4836	-16.1654	0.9028	348.5534	-25.2993	0.8998	329.1911	-23.9584	0.9169
	a_1	349.4886	-4.5124	-0.0022	348.0500	-1.9937	0.0023	347.8657	2.6929	0.0074
	a_2	-0.2459	0.3733		-0.1824	0.5932		0.2083	0.5382	
12	a_0	341.7259	-20.3026	0.9006	336.4234	-26.6990	0.9021	317.2651	-20.7291	0.9244
	a_1	348.9904	-3.7650	-0.0006	347.6815	-0.8019	0.0039	348.2915	3.7712	0.0087
	a_2	-0.2677	0.4635		-0.0668	0.6313		0.2062	0.4504	
13	a_0	330.4491	-23.6023	0.8999	324.0406	-26.8696	0.9060	305.7616	-16.5095	0.9331
	a_1	348.4476	-2.8358	0.0012	347.5499	0.4665	0.0057	348.7110	4.6727	0.0100
	a_2	-0.2415	0.5446		0.0561	0.6308		0.1472	0.3530	
14	a_0	318.6567	-25.8922	0.9012	311.6485	-25.7730	0.9117	294.6177	-11.4859	0.9432
	a_1	347.9581	-1.7428	0.0034	347.6690	1.7328	0.0077	349.0105	5.3795	0.0113
	a_2	-0.1665	0.6066		0.1482	0.5939		0.0457	0.2500	
15	a_0	306.4505	-27.0277	0.9045	299.4665	-23.4477	0.9194	283.6712	-5.8588	0.9544
	a_1	347.6226	-0.5247	0.0057	347.9744	2.9236	0.0097	349.1053	5.8812	0.0122
	a_2	-0.0588	0.6396		0.1846	0.5304		-0.0856	0.1375	

$GHA^\circ, DEC^\circ = (a_2 x + a_1)x + a_0$ $HP^\circ = a_1 x + a_0$ where $x = GMT^h/24$

Day		April GHA °	April DEC °	April HP °	May GHA °	May DEC °	May HP °	June GHA °	June DEC °	June HP °
16	a_0	294.0167	-26.9127	0.9103	287.6250	-19.9954	0.9292	272.6879	0.1572	0.9667
	a_1	347.5078	0.7596	0.0082	348.3519	3.9860	0.0117	348.9359	6.1598	0.0126
	a_2	0.0499	0.6385		0.1608	0.4506		-0.2358	0.0061	
17	a_0	281.5759	-25.5153	0.9185	276.1362	-15.5606	0.9410	261.3847	6.3198	0.9793
	a_1	347.6144	2.0409	0.0108	348.6800	4.8887	0.0135	348.4631	6.1772	0.0122
	a_2	0.1258	0.6051		0.0857	0.3602		-0.3906	-0.1565	
18	a_0	269.3165	-22.8705	0.9293	264.8995	-10.3140	0.9545	249.4545	12.3363	0.9916
	a_1	347.8744	3.2544	0.0131	348.8560	5.6112	0.0148	347.6749	5.8698	0.0108
	a_2	0.1486	0.5457		-0.0278	0.2567		-0.5208	-0.3587	
19	a_0	257.3389	-19.0720	0.9424	253.7249	-4.4487	0.9693	236.6076	17.8426	1.0025
	a_1	348.1794	4.3485	0.0149	348.8033	6.1285	0.0152	346.6165	5.1550	0.0082
	a_2	0.1153	0.4655		-0.1679	0.1313		-0.5720	-0.5929	
20	a_0	245.6318	-14.2602	0.9574	242.3571	1.8078	0.9846	222.6542	22.4000	1.0107
	a_1	348.4158	5.2826	0.0160	348.4681	6.3968	0.0144	345.4464	3.9634	0.0044
	a_2	0.0343	0.3647		-0.3217	-0.0288		-0.4739	-0.8207	
21	a_0	234.0796	-8.6155	0.9734	230.5005	8.1715	0.9990	207.6324	25.5397	1.0151
	a_1	348.4885	6.0164	0.0158	347.8206	6.3457	0.0121	344.4754	2.3049	-0.0002
	a_2	-0.0817	0.2375		-0.4661	-0.2338		-0.1970	-0.9717	
22	a_0	222.4836	-2.3649	0.9893	217.8531	14.2782	1.0112	191.9183	26.8727	1.0149
	a_1	348.3269	6.4971	0.0142	346.8756	5.8825	0.0084	344.0793	0.3383	-0.0051
	a_2	-0.2189	0.0746		-0.5560	-0.4806		0.1756	-0.9807	
23	a_0	210.5887	4.2026	1.0036	204.1734	19.6749	1.0196	176.1795	26.2331	1.0098
	a_1	347.8872	6.6524	0.0110	345.7401	4.9185	0.0034	344.4517	-1.6409	-0.0094
	a_2	-0.3580	-0.1308		-0.5216	-0.7376		0.4813	-0.8434	
24	a_0	198.1157	10.7191	1.0147	189.3964	23.8516	1.0230	161.1154	23.7533	1.0003
	a_1	347.1629	6.3950	0.0065	344.6701	3.4288	-0.0021	345.4412	-3.3337	-0.0128
	a_2	-0.4646	-0.3748		-0.3069	-0.9387		0.6152	-0.6216	
25	a_0	184.8136	16.7340	1.0212	173.7669	26.3403	1.0209	147.1714	19.8028	0.9875
	a_1	346.2161	5.6439	0.0011	344.0445	1.5281	-0.0074	346.6901	-4.5738	-0.0148
	a_2	-0.4838	-0.6325		0.0514	-1.0085		0.5904	-0.3903	
26	a_0	170.5486	21.7409	1.0223	157.8701	26.8618	1.0135	134.4496	14.8428	0.9726
	a_1	345.2247	4.3683	-0.0043	344.1616	-0.5099	-0.0117	347.8789	-5.3476	-0.0153
	a_2	-0.3569	-0.8525		0.4040	-0.9203		0.4750	-0.1938	
27	a_0	155.4221	25.2543	1.0180	142.4400	25.4355	1.0017	122.8004	9.3045	0.9572
	a_1	344.4934	2.6441	-0.0090	344.9972	-2.3604	-0.0146	348.8305	-5.7286	-0.0147
	a_2	-0.0789	-0.9700		0.6070	-0.7225		0.3270	-0.0430	
28	a_0	139.8434	26.9288	1.0089	128.0448	22.3573	0.9871	111.9548	3.5352	0.9425
	a_1	344.3385	0.6834	-0.0124	346.2339	-3.8044	-0.0159	349.4832	-5.8099	-0.0131
	a_2	0.2554	-0.9469		0.6326	-0.4947		0.1776	0.0699	
29	a_0	124.4425	26.6681	0.9964	114.9095	18.0624	0.9711	101.6127	-2.2029	0.9294
	a_1	344.8702	-1.2236	-0.0144	347.5108	-4.7876	-0.0158	349.8362	-5.6674	-0.0108
	a_2	0.5054	-0.8045		0.5403	-0.2926		0.0392	0.1586	
30	a_0	109.8201	24.6441	0.9819	102.9577	12.9855	0.9553	91.4855	-7.7100	0.9185
	a_1	345.9043	-2.8354	-0.0149	348.5949	-5.3661	-0.0146	349.9117	-5.3493	-0.0082
	a_2	0.5988	-0.6062		0.3975	-0.1343		-0.0830	0.2363	
31	a_0				91.9469	7.4876	0.9407			
	a_1				349.3893	-5.6298	-0.0127			
	a_2				0.2452	-0.0151				

Table 2: MOON, 1990

Day		July GHA °	July DEC °	July HP °	August GHA °	August DEC °	August HP °	September GHA °	September DEC °	September HP °
1	a_0	81.3121	-12.8214	0.9103	67.2737	-25.7267	0.9039	48.9449	-23.6688	0.9244
	a_1	349.7419	-4.8770	-0.0055	347.9410	-1.7282	0.0030	347.7601	2.8401	0.0105
	a_2	-0.1839	0.3132		-0.1846	0.6040		0.1539	0.5605	
2	a_0	70.8687	-17.3835	0.9048	55.0323	-26.8501	0.9070	36.8590	-20.2699	0.9349
	a_1	349.3686	-4.2513	-0.0029	347.5671	-0.5154	0.0052	348.0754	3.9648	0.0114
	a_2	-0.2539	0.3946		-0.0761	0.6427		0.1574	0.4769	
3	a_0	59.9830	-21.2385	0.9019	42.5258	-26.7228	0.9122	25.0909	-15.8305	0.9464
	a_1	348.8532	-3.4621	-0.0004	347.4156	0.7760	0.0070	348.3967	4.9216	0.0115
	a_2	-0.2790	0.4797		0.0474	0.6437		0.1123	0.3694	
4	a_0	48.5579	-24.2192	0.9016	29.9909	-25.3039	0.9193	13.5982	-10.5421	0.9580
	a_1	348.2862	-2.5009	0.0018	347.5166	2.0692	0.0083	348.6258	5.6632	0.0108
	a_2	-0.2461	0.5603		0.1475	0.6032		0.0290	0.2410	
5	a_0	36.5998	-26.1586	0.9034	17.6561	-22.6330	0.9276	2.2507	-4.6410	0.9688
	a_1	347.7862	-1.3763	0.0037	347.8203	3.2802	0.0090	348.6864	6.1479	0.0092
	a_2	-0.1538	0.6214		0.1963	0.5264		-0.0797	0.0928	
6	a_0	24.2350	-26.9129	0.9071	5.6725	-18.8286	0.9367	350.8549	1.5963	0.9780
	a_1	347.4758	-0.1273	0.0052	348.2212	4.3359	0.0092	348.5274	6.3360	0.0071
	a_2	-0.0229	0.6474		0.1864	0.4232		-0.2002	-0.0744	
7	a_0	11.6906	-26.3932	0.9123	354.0789	-14.0720	0.9459	339.1796	7.8540	0.9851
	a_1	347.4337	1.1740	0.0063	348.6003	5.1841	0.0089	348.1241	6.1889	0.0047
	a_2	0.1067	0.6291		0.1258	0.3031		-0.3141	-0.2584	
8	a_0	359.2328	-24.5914	0.9186	342.8030	-8.5875	0.9548	326.9880	13.7805	0.9898
	a_1	347.6553	2.4374	0.0072	348.8561	5.7916	0.0082	347.4880	5.6724	0.0022
	a_2	0.1949	0.5692		0.0286	0.1719		-0.3931	-0.4521	
9	a_0	347.0836	-21.5866	0.9258	331.6851	-2.6269	0.9630	314.0829	18.9969	0.9920
	a_1	348.0547	3.5791	0.0078	348.9155	6.1367	0.0073	346.6879	4.7652	-0.0001
	a_2	0.2211	0.4794		-0.0919	0.0309		-0.3976	-0.6395	
10	a_0	335.3587	-17.5303	0.9336	320.5059	3.5374	0.9703	300.3754	23.1195	0.9919
	a_1	348.5049	4.5394	0.0083	348.7319	6.2001	0.0063	345.8755	3.4783	-0.0019
	a_2	0.1864	0.3730		-0.2226	-0.1217		-0.2924	-0.7916	
11	a_0	324.0483	-12.6202	0.9419	309.0127	9.6123	0.9765	285.9629	25.8045	0.9899
	a_1	348.8833	5.2861	0.0087	348.2838	5.9587	0.0052	345.2789	1.8822	-0.0034
	a_2	0.1040	0.2590		-0.3458	-0.2886		-0.0811	-0.8727	
12	a_0	313.0332	-7.0775	0.9506	296.9488	15.2787	0.9817	271.1658	26.8143	0.9865
	a_1	349.0950	5.8048	0.0089	347.5838	5.3829	0.0041	345.1182	0.1226	-0.0046
	a_2	-0.0115	0.1396		-0.4323	-0.4690		0.1704	-0.8585	
13	a_0	302.1138	-1.1358	0.9595	284.1003	20.1887	0.9858	256.4584	26.0805	0.9819
	a_1	349.0741	6.0857	0.0091	346.7041	4.4432	0.0029	345.4728	-1.6053	-0.0056
	a_2	-0.1480	0.0111		-0.4389	-0.6515		0.3682	-0.7558	
14	a_0	291.0369	4.9579	0.9686	270.3678	23.9773	0.9887	242.3013	23.7226	0.9763
	a_1	348.7781	6.1107	0.0090	345.8072	3.1331	0.0015	346.2263	-3.1219	-0.0064
	a_2	-0.2935	-0.1344		-0.3253	-0.8066		0.4568	-0.5978	
15	a_0	279.5188	10.9307	0.9776	255.8545	26.3020	0.9902	228.9842	20.0066	0.9698
	a_1	348.1873	5.8457	0.0085	345.1434	1.5064	-0.0001	347.1526	-4.3176	-0.0072
	a_2	-0.4280	-0.3049		-0.0942	-0.8918		0.4415	-0.4219	

GHA°, DEC° = $(a_2 x + a_1)x + a_0$ HP° = $a_1 x + a_0$ where $x = GMT^h/24$

Day		July GHA °	July DEC °	July HP °	August GHA °	August DEC °	August HP °	September GHA °	September DEC °	September HP °
16	a_0	267.2763	16.4675	0.9862	240.9094	26.9169	0.9900	216.5766	15.2705	0.9626
	a_1	347.3204	5.2389	0.0075	344.9571	-0.2930	-0.0021	348.0422	-5.1591	-0.0080
	a_2	-0.5142	-0.5015		0.1803	-0.8743		0.3598	-0.2531	
17	a_0	254.0830	21.2006	0.9936	226.0511	25.7520	0.9880	204.9762	9.8615	0.9546
	a_1	346.2724	4.2344	0.0057	345.3334	-2.0539	-0.0042	348.7638	-5.6622	-0.0085
	a_2	-0.4958	-0.7068		0.3923	-0.7573		0.2481	-0.1017	
18	a_0	239.8631	24.7248	0.9994	211.7788	22.9445	0.9838	193.9857	4.1002	0.9461
	a_1	345.2577	2.8110	0.0031	346.1368	-3.5739	-0.0064	349.2597	-5.8629	-0.0088
	a_2	-0.3255	-0.8761		0.4823	-0.5780		0.1302	0.0307	
19	a_0	224.8016	26.6582	1.0025	198.3975	18.7967	0.9774	183.3733	-1.7296	0.9372
	a_1	344.5944	1.0408	-0.0001	347.1148	-4.7294	-0.0084	349.5184	-5.7993	-0.0088
	a_2	-0.0244	-0.9497		0.4605	-0.3808		0.0188	0.1467	
20	a_0	209.3781	26.7505	1.0024	185.9709	13.6902	0.9689	172.9084	-7.3800	0.9284
	a_1	344.5542	-0.8776	-0.0037	348.0423	-5.4875	-0.0100	349.5533	-5.5040	-0.0082
	a_2	0.2934	-0.8921		0.3700	-0.1968		-0.0786	0.2500	
21	a_0	194.2300	24.9843	0.9987	174.3808	8.0093	0.9589	162.3815	-12.6320	0.9202
	a_1	345.1633	-2.6736	-0.0073	348.7842	-5.8766	-0.0109	349.3923	-5.0026	-0.0072
	a_2	0.5010	-0.7244		0.2506	-0.0391		-0.1548	0.3436	
22	a_0	179.8955	21.5907	0.9914	163.4130	2.0962	0.9479	151.6180	-17.2892	0.9130
	a_1	346.1865	-4.1249	-0.0103	349.2848	-5.9509	-0.0111	349.0776	-4.3141	-0.0056
	a_2	0.5552	-0.5061		0.1266	0.0908		-0.2011	0.4287	
23	a_0	166.6360	16.9641	0.9810	152.8219	-3.7616	0.9368	140.4944	-21.1730	0.9073
	a_1	347.3094	-5.1333	-0.0124	349.5361	-5.7663	-0.0104	348.6692	-3.4548	-0.0036
	a_2	0.4932	-0.2922		0.0102	0.1980		-0.2083	0.5041	
24	a_0	154.4361	11.5424	0.9686	142.3661	-9.3280	0.9264	128.9560	-24.1224	0.9038
	a_1	348.3005	-5.7117	-0.0134	349.5537	-5.3684	-0.0090	348.2462	-2.4439	-0.0011
	a_2	0.3725	-0.1115		-0.0915	0.2895		-0.1714	0.5652	
25	a_0	143.1062	5.7222	0.9551	131.8266	-14.4051	0.9173	117.0323	-26.0003	0.9026
	a_1	349.0459	-5.9292	-0.0133	349.3668	-4.7883	-0.0070	347.8988	-1.3099	0.0016
	a_2	0.2340	0.0302		-0.1712	0.3720		-0.0960	0.6053	
26	a_0	132.3833	-0.1745	0.9417	121.0212	-18.8198	0.9103	104.8370	-26.7046	0.9042
	a_1	349.5125	-5.8647	-0.0122	349.0190	-4.0435	-0.0046	347.7062	-0.0951	0.0044
	a_2	0.0985	0.1400		-0.2196	0.4495		-0.0016	0.6188	
27	a_0	121.9918	-5.8974	0.9295	109.8204	-22.4123	0.9057	92.5434	-26.1812	0.9087
	a_1	349.7072	-5.5824	-0.0104	348.5729	-3.1434	-0.0018	347.7065	1.1466	0.0073
	a_2	-0.0250	0.2289		-0.2261	0.5216		0.0836	0.6037	
28	a_0	111.6718	-11.2492	0.9191	98.1682	-25.0327	0.9039	80.3345	-24.4317	0.9160
	a_1	349.6541	-5.1237	-0.0080	348.1137	-2.0979	0.0011	347.8800	2.3576	0.0099
	a_2	-0.1306	0.3080		-0.1842	0.5830		0.1345	0.5624	
29	a_0	101.1935	-16.0633	0.9111	86.0994	-26.5468	0.9050	68.3491	-21.5130	0.9259
	a_1	349.3882	-4.5078	-0.0052	347.7404	-0.9282	0.0039	348.1560	3.4853	0.0121
	a_2	-0.2109	0.3850		-0.0997	0.6246		0.1387	0.4989	
30	a_0	90.3700	-20.1844	0.9059	73.7422	-26.8502	0.9090	56.6430	-17.5306	0.9381
	a_1	348.9600	-3.7378	-0.0024	347.5403	0.3257	0.0066	348.4395	4.4858	0.0136
	a_2	-0.2541	0.4633		0.0053	0.6370		0.0967	0.4156	
31	a_0	79.0760	-23.4573	0.9035	61.2898	-25.8879	0.9156			
	a_1	348.4438	-2.8103	0.0004	347.5550	1.6048	0.0088			
	a_2	-0.2474	0.5396		0.0990	0.6154				

Table 2: MOON, 1990

Day		October GHA °	October DEC °	October HP °	November GHA °	November DEC °	November HP °	December GHA °	December DEC °	December HP °
1	a_0	45.1775	-12.6313	0.9517	31.0094	8.9940	0.9977	23.8519	22.2430	1.0194
	a_1	348.6377	5.3202	0.0141	347.6964	6.1617	0.0117	345.1044	3.8058	0.0054
	a_2	0.0168	0.3111		-0.4698	-0.2660		-0.4661	-0.8489	
2	a_0	33.8297	-7.0027	0.9659	18.2348	14.8846	1.0094	8.4967	25.1969	1.0248
	a_1	348.6743	5.9464	0.0135	346.7432	5.6324	0.0075	344.1505	2.0880	-0.0003
	a_2	-0.0905	0.1809		-0.5298	-0.5110		-0.1519	-0.9934	
3	a_0	22.4109	-0.8788	0.9795	4.4496	20.0011	1.0169	352.5032	26.2917	1.0245
	a_1	348.4942	6.3127	0.0117	345.6610	4.6054	0.0024	343.8495	0.0765	-0.0059
	a_2	-0.2131	0.0198		-0.4646	-0.7519		0.2360	-0.9798	
4	a_0	10.6895	5.4496	0.9913	349.6507	23.8511	1.0194	336.5947	25.3918	1.0185
	a_1	348.0658	6.3565	0.0088	344.7093	3.0862	-0.0027	344.3464	-1.8998	-0.0107
	a_2	-0.3334	-0.1735		-0.2373	-0.9243		0.5258	-0.8185	
5	a_0	358.4201	11.6280	1.0001	334.1295	26.0121	1.0166	321.4691	22.6783	1.0077
	a_1	347.3914	6.0116	0.0051	344.2283	1.2161	-0.0073	345.4252	-3.5407	-0.0140
	a_2	-0.4217	-0.3918		0.0984	-0.9672		0.6297	-0.5844	
6	a_0	345.3896	17.2433	1.0052	318.4625	26.2629	1.0092	307.5232	18.5580	0.9936
	a_1	346.5332	5.2257	0.0011	344.4406	-0.7362	-0.0108	346.7017	-4.7049	-0.0157
	a_2	-0.4338	-0.6132		0.4040	-0.8703		0.5799	-0.3532	
7	a_0	331.4912	21.8519	1.0062	303.3105	24.6602	0.9983	294.8021	13.5040	0.9778
	a_1	345.6462	3.9903	-0.0027	345.2733	-2.4846	-0.0129	347.8679	-5.4039	-0.0159
	a_2	-0.3258	-0.7989		0.5662	-0.6835		0.4502	-0.1628	
8	a_0	316.8163	25.0412	1.0035	289.1502	21.4964	0.9854	283.1171	7.9402	0.9618
	a_1	344.9804	2.3772	-0.0058	346.4251	-3.8504	-0.0137	348.7690	-5.7234	-0.0150
	a_2	-0.0956	-0.9019		0.5734	-0.4745		0.2968	-0.0182	
9	a_0	301.7069	26.5167	0.9977	276.1469	17.1753	0.9717	272.1798	2.2010	0.9468
	a_1	344.7904	0.5567	-0.0080	347.5817	-4.7944	-0.0134	349.3606	-5.7556	-0.0133
	a_2	0.1847	-0.8925		0.4805	-0.2874		0.1476	0.0919	
10	a_0	286.6866	26.1831	0.9897	264.2062	12.0967	0.9583	261.6852	-3.4608	0.9335
	a_1	345.1757	-1.2402	-0.0093	348.5454	-5.3640	-0.0124	349.6529	-5.5697	-0.0111
	a_2	0.4059	-0.7822		0.3449	-0.1355		0.0140	0.1817	
11	a_0	272.2703	24.1642	0.9804	253.0936	6.5998	0.9458	251.3497	-8.8470	0.9224
	a_1	346.0068	-2.8089	-0.0098	349.2346	-5.6312	-0.0110	349.6772	-5.2056	-0.0089
	a_2	0.5029	-0.6147		0.2021	-0.0140		-0.0981	0.2634	
12	a_0	258.7796	20.7442	0.9705	242.5275	0.9568	0.9348	240.9271	-13.7876	0.9135
	a_1	347.0266	-4.0373	-0.0098	349.6367	-5.6569	-0.0095	349.4761	-4.6789	-0.0066
	a_2	0.4836	-0.4357		0.0684	0.0881		-0.1818	0.3442	
13	a_0	246.2880	16.2747	0.9607	232.2301	-4.6101	0.9253	230.2204	-18.1206	0.9069
	a_1	348.0007	-4.9053	-0.0095	349.7705	-5.4799	-0.0079	349.1059	-3.9903	-0.0045
	a_2	0.3926	-0.2715		-0.0489	0.1808		-0.2268	0.4260	
14	a_0	234.6788	11.1008	0.9512	221.9498	-9.9073	0.9174	219.0996	-21.6833	0.9024
	a_1	348.7879	-5.4450	-0.0090	349.6689	-5.1181	-0.0064	348.6443	-3.1370	-0.0027
	a_2	0.2714	-0.1296		-0.1438	0.2712		-0.2214	0.5037	
15	a_0	223.7356	5.5289	0.9422	211.4736	-14.7523	0.9110	207.5239	-24.3153	0.8997
	a_1	349.3302	-5.7016	-0.0084	349.3760	-4.5755	-0.0050	348.1939	-2.1264	-0.0010
	a_2	0.1455	-0.0067		-0.2080	0.3618		-0.1605	0.5664	

$GHA^\circ, DEC^\circ = (a_2 x + a_1)x + a_0$ $HP^\circ = a_1 x + a_0$ where $x = GMT^h/24$

Day		October			November			December		
		GHA °	DEC °	HP °	GHA °	DEC °	HP °	GHA °	DEC °	HP °
16	a_0	213.2087	-0.1772	0.9337	200.6412	-18.9642	0.9059	195.5594	-25.8746	0.8988
	a_1	349.6191	-5.7134	-0.0078	348.9531	-3.8509	-0.0037	347.8681	-0.9885	0.0005
	a_2	0.0276	0.1032		-0.2304	0.4497		-0.0552	0.6010	
17	a_0	202.8533	-5.7853	0.9259	189.3644	-22.3638	0.9022	183.3748	-26.2621	0.8993
	a_1	349.6715	-5.5062	-0.0071	348.4845	-2.9490	-0.0024	347.7584	0.2192	0.0019
	a_2	-0.0750	0.2054		-0.2017	0.5267		0.0652	0.5985	
18	a_0	192.4481	-11.0840	0.9188	177.6488	-24.7849	0.8998	171.2004	-25.4451	0.9012
	a_1	349.5177	-5.0946	-0.0063	348.0747	-1.8914	-0.0010	347.8945	1.4211	0.0033
	a_2	-0.1554	0.3031		-0.1229	0.5813		0.1640	0.5591	
19	a_0	181.8093	-15.8735	0.9125	165.6029	-26.0945	0.8989	159.2600	-23.4662	0.9045
	a_1	349.2015	-4.4873	-0.0053	347.8266	-0.7235	0.0006	348.2306	2.5424	0.0047
	a_2	-0.2050	0.3959		-0.0124	0.6033		0.2144	0.4921	
20	a_0	170.8057	-19.9632	0.9072	153.4194	-26.2151	0.8995	147.7049	-20.4334	0.9092
	a_1	348.7850	-3.6937	-0.0040	347.8047	0.4881	0.0024	348.6673	3.5278	0.0063
	a_2	-0.2137	0.4800		0.0972	0.5891		0.2088	0.4100	
21	a_0	159.3778	-23.1756	0.9032	141.3229	-25.1388	0.9019	136.5799	-16.4975	0.9154
	a_1	348.3508	-2.7308	-0.0023	348.0059	1.6701	0.0044	349.0911	4.3479	0.0080
	a_2	-0.1758	0.5483		0.1731	0.5444		0.1539	0.3228	
22	a_0	147.5544	-25.3571	0.9009	129.5024	-22.9257	0.9063	125.8229	-11.8286	0.9234
	a_1	347.9944	-1.6301	-0.0004	348.3602	2.7607	0.0067	349.4034	4.9936	0.0097
	a_2	-0.0964	0.5924		0.1969	0.4798		0.0616	0.2345	
23	a_0	135.4545	-26.3945	0.9005	118.0589	-19.6867	0.9130	115.2854	-6.6025	0.9331
	a_1	347.8008	-0.4406	0.0020	348.7613	3.7210	0.0091	349.5298	5.4635	0.0115
	a_2	0.0041	0.6063		0.1676	0.4060		-0.0581	0.1424	
24	a_0	123.2612	-26.2292	0.9024	106.9863	-15.5613	0.9221	104.7541	-0.9988	0.9447
	a_1	347.8126	0.7761	0.0045	349.1020	4.5334	0.0115	349.4157	5.7507	0.0131
	a_2	0.0955	0.5896		0.0939	0.3280		-0.1979	0.0382	
25	a_0	111.1705	-24.8643	0.9070	96.1800	-10.7016	0.9335	93.9688	4.7874	0.9578
	a_1	348.0102	1.9584	0.0072	349.2939	5.1907	0.0137	349.0204	5.8315	0.0142
	a_2	0.1512	0.5480		-0.0134	0.2443		-0.3494	-0.0903	
26	a_0	99.3321	-22.3591	0.9142	85.4577	-5.2687	0.9472	82.6368	10.5251	0.9720
	a_1	348.3201	3.0564	0.0100	349.2701	5.6820	0.0154	348.3184	5.6569	0.0145
	a_2	0.1582	0.4894		-0.1457	0.1466		-0.4945	-0.2558	
27	a_0	87.8095	-18.8148	0.9242	74.5790	0.5574	0.9627	70.4587	15.9219	0.9865
	a_1	348.6429	4.0366	0.0125	348.9802	5.9801	0.0162	347.3184	5.1508	0.0136
	a_2	0.1171	0.4193		-0.2951	0.0220		-0.5939	-0.4634	
28	a_0	76.5678	-14.3607	0.9367	63.2610	6.5560	0.9790	57.1835	20.6045	1.0002
	a_1	348.8820	4.8769	0.0145	348.3885	6.0308	0.0159	346.1089	4.2244	0.0113
	a_2	0.0361	0.3376		-0.4469	-0.1439		-0.5821	-0.6956	
29	a_0	65.4835	-9.1482	0.9512	51.2001	12.4384	0.9949	42.7142	24.1291	1.0116
	a_1	348.9578	5.5552	0.0158	347.4866	5.7497	0.0139	344.9177	2.8230	0.0076
	a_2	-0.0750	0.2384		-0.5685	-0.3590		-0.3968	-0.8986	
30	a_0	54.3635	-3.3572	0.9671	38.1176	17.8240	1.0089	27.2422	26.0515	1.0192
	a_1	348.8098	6.0369	0.0159	346.3307	5.0343	0.0104	344.1089	1.0046	0.0027
	a_2	-0.2070	0.1113		-0.5992	-0.6104		-0.0549	-0.9940	
31	a_0	42.9634	2.7876	0.9830				11.3037	26.0634	1.0219
	a_1	348.3954	6.2657	0.0145				344.0093	-1.0063	-0.0028
	a_2	-0.3468	-0.0550					0.3093	-0.9325	

Table 3: Moon Monthly Chebyshev Coefficients, 1986

	January			February			March		
	λ °	β °	HP °	λ °	β °	HP °	λ °	β °	HP °
a_0	356.4283	1.3937	0.9572	46.3859	-0.4032	0.9662	55.9280	-0.8213	0.9688
a_1	211.0455	0.3577	-0.0055	210.6941	0.4036	-0.0003	210.3689	0.4642	0.0042
a_2	-4.3390	2.6861	0.0117	-0.8988	-1.1247	0.0313	0.6590	-2.0247	0.0380
a_3	1.7440	-3.2312	0.0348	3.9738	-4.1002	0.0078	4.4817	-3.9334	-0.0018
a_4	2.6391	-1.4102	-0.0118	0.6410	0.5873	-0.0229	0.2442	0.9974	-0.0229
a_5	-0.8541	0.8488	-0.0118	-1.4402	0.8566	-0.0035	-1.2718	0.8057	-0.0052
a_6	-0.6532	0.0386	0.0053	-0.1387	-0.0790	0.0061	-0.2980	-0.0564	0.0038
a_7	0.2916	-0.1081	0.0021	0.3139	0.0388		0.1762	0.0409	
a_8	0.0972	0.0826		0.0025	0.0044		0.0684	-0.0126	
a_9	-0.0775	0.0049		-0.0304	-0.0444		-0.0064	-0.0400	
a_{10}	0.0092	-0.0295		0.0086	0.0008		0.0014	-0.0034	
a_{11}	0.0155	0.0054		-0.0077	0.0093		-0.0044	0.0071	
a_{12}	-0.0103	0.0056		-0.0027			-0.0037		
a_{13}	-0.0014	-0.0032		0.0036					
a_{14}	0.0033								
a_{15}									
sum	566.3382	0.6412	0.9820	259.5049	-3.8507	0.9847	270.3435	-4.5765	0.9849
n	15	14	8	14	12	7	13	12	7

	April			May			June		
	λ °	β °	HP °	λ °	β °	HP °	λ °	β °	HP °
a_0	106.2375	-1.9542	0.9632	142.3403	-1.7477	0.9542	190.5334	-0.3757	0.9395
a_1	210.5045	0.2233	0.0094	210.6705	-0.0732	0.0125	211.3025	-0.4790	0.0121
a_2	4.7322	-4.5209	0.0266	6.5504	-4.2225	0.0078	5.9232	-1.3884	-0.0245
a_3	3.4819	-1.0897	-0.0335	1.3655	1.5787	-0.0466	-2.4495	3.8568	-0.0422
a_4	-1.9978	2.3368	-0.0201	-3.0216	2.3053	-0.0110	-3.1178	0.8747	0.0074
a_5	-1.2347	0.4080	0.0032	-0.9066	-0.1416	0.0086	-0.0018	-0.9689	0.0140
a_6	0.0089	-0.1572	0.0069	0.1932	-0.3550	0.0077	0.6621	-0.3387	0.0040
a_7	0.2482	-0.1046	0.0023	0.3415	-0.1428	0.0009	0.3285	0.0887	-0.0034
a_8	0.0926	-0.0715		0.0968	0.0027	-0.0022	-0.0917	0.1151	-0.0025
a_9	-0.0147	0.0024		-0.0509	0.0482		-0.1216	0.0210	
a_{10}	-0.0206	0.0232		-0.0381	0.0228		-0.0112	-0.0264	
a_{11}	-0.0082	0.0089		-0.0065	-0.0041		0.0272	-0.0171	
a_{12}	0.0007	-0.0020		0.0062	-0.0088		0.0126	0.0024	
a_{13}	0.0025	-0.0029		0.0048	-0.0025		-0.0029	0.0063	
a_{14}					0.0013		-0.0047	0.0014	
a_{15}									
sum	322.0330	-4.9004	0.9580	357.5455	-2.7392	0.9319	402.9883	1.3722	0.9044
n	14	14	8	14	15	9	15	15	9

Evaluate the recurrence relation :

$b_i = 2(2x-1)b_{i+1} - b_{i+2} + a_i$ for $i=n-1,n-2,\ldots 2,1,0$

where $b_{n+1} = b_n = 0$ and $x = (d + GMT^h/24)/32$

Then $\lambda°, \beta°, HP° = (b_0 - b_2 + a_0)/2$

	July			August			September		
	λ °	β °	HP °	λ °	β °	HP °	λ °	β °	HP °
a_0	224.8913	0.8819	0.9332	271.5789	2.0466	0.9323	318.6351	2.0121	0.9420
a_1	211.5262	-0.5135	0.0088	211.5669	-0.1366	0.0018	210.9416	0.4228	-0.0035
a_2	3.7051	1.3366	-0.0382	-0.4952	3.9341	-0.0401	-3.8699	3.8997	-0.0178
a_3	-4.1601	3.7604	-0.0263	-4.5216	1.4407	0.0047	-2.1719	-1.9696	0.0309
a_4	-2.1000	-0.6946	0.0172	0.0648	-2.2453	0.0216	1.9209	-2.1167	0.0119
a_5	0.6559	-1.1130	0.0112	1.1742	-0.3999	0.0029	0.7153	0.6373	-0.0042
a_6	0.6888	0.0306	0.0000	0.3255	0.4822	-0.0038	-0.1264	0.3075	-0.0027
a_7	0.1316	0.2310	-0.0040	-0.1916	0.0771	-0.0025	-0.1625	-0.1714	
a_8	-0.1939	0.0466		-0.1556	-0.0995		-0.0167	-0.0313	
a_9	-0.0855	-0.0441		0.0409	-0.0269		0.0429	0.0369	
a_{10}	0.0371	-0.0270		0.0463	0.0192		-0.0008	0.0068	
a_{11}	0.0307	0.0042		-0.0093	0.0111		-0.0113	-0.0056	
a_{12}	-0.0026	0.0101		-0.0126	-0.0031		0.0018	-0.0025	
a_{13}	-0.0089	0.0015		0.0015	-0.0040		0.0025		
a_{14}	-0.0018	-0.0028		0.0034					
a_{15}	0.0021								
sum	435.1160	3.9079	0.9019	479.4165	5.0957	0.9169	525.9006	3.0260	0.9566
n	16	15	8	15	14	8	14	13	7

	October			November			December		
	λ °	β °	HP °	λ °	β °	HP °	λ °	β °	HP °
a_0	353.4507	1.1788	0.9536	42.5276	-0.5537	0.9702	78.8258	-1.5299	0.9770
a_1	210.1379	0.6619	-0.0054	209.3364	0.5748	-0.0027	209.0351	0.2818	-0.0012
a_2	-5.0678	2.0835	0.0088	-3.6370	-1.7185	0.0457	-1.3260	-3.8712	0.0608
a_3	0.6654	-3.8507	0.0398	4.5001	-4.0199	0.0290	6.1218	-2.1090	0.0101
a_4	2.4185	-0.9964	-0.0010	1.6995	0.9848	-0.0158	0.5418	2.0176	-0.0216
a_5	-0.0152	0.9757	-0.0045	-0.5218	0.7169	-0.0024	-0.5806	0.3284	-0.0002
a_6	-0.0912	-0.0354	-0.0007	0.1018	-0.1933	-0.0026	0.1268	-0.1092	-0.0040
a_7	-0.0175	-0.1296		-0.1402	0.0596		-0.2894	0.0566	
a_8	-0.0523	0.0559		-0.0878	0.0243		-0.0515	-0.0347	
a_9	0.0109	0.0113		0.0391	-0.0147		0.0514	-0.0049	
a_{10}	0.0067	-0.0058		-0.0006	0.0023		-0.0042	-0.0032	
a_{11}	-0.0049			-0.0038	-0.0047		0.0049	-0.0054	
a_{12}				0.0037			0.0023	0.0030	
a_{13}									
a_{14}									
a_{15}									
sum	561.4412	-0.0508	0.9906	253.8170	-4.1421	1.0214	292.4582	-4.9801	1.0209
n	12	11	7	13	12	7	13	13	7

Calculate $X = \cos\beta \cos\lambda$ $Y = \cos\epsilon \cos\beta \sin\lambda - \sin\epsilon \sin\beta$

and $Z = \sin\epsilon \cos\beta \sin\lambda + \cos\epsilon \sin\beta$ where $\epsilon = 23^\circ.44$

Then $RA^\circ = \tan^{-1}(Y/X)$ and $DEC^\circ = \sin^{-1}(Z)$

Add 180° to RA if $X < 0$. Add 360° to RA if $X > 0$ and $Y < 0$.

	January			February			March		
	λ °	β °	HP °	λ °	β °	HP °	λ °	β °	HP °
a_0	129.2316	-1.5946	0.9738	178.6928	-0.2290	0.9587	187.2954	0.0729	0.9563
a_1	209.1482	-0.1680	0.0034	209.7492	-0.5751	0.0067	210.2142	-0.6656	0.0031
a_2	3.1055	-4.0272	0.0539	5.4485	-1.0541	0.0207	4.5688	-0.3613	0.0138
a_3	5.4333	1.6952	-0.0236	1.9478	4.0907	-0.0413	1.0972	4.2958	-0.0372
a_4	-1.2747	2.1252	-0.0193	-2.3833	0.7133	-0.0070	-2.3810	0.3051	-0.0020
a_5	-0.5445	-0.2085	0.0003	-0.3467	-0.8451	0.0031	0.0227	-0.9544	0.0061
a_6	-0.1999	-0.1582	-0.0029	-0.1243	-0.2625	0.0002	0.1455	-0.2092	-0.0023
a_7	-0.2464	-0.0813	0.0020	0.0231	0.0020	0.0025	-0.0769	0.0566	
a_8	0.0899	-0.0276		0.1350	0.0567		0.0357	0.0703	
a_9	0.0561	0.0062		0.0208	0.0159		0.0200	-0.0002	
a_{10}	0.0074	-0.0001		-0.0182	0.0025		-0.0018	-0.0092	
a_{11}	-0.0001	0.0065		-0.0117	-0.0003		-0.0051	0.0015	
a_{12}	-0.0045	0.0025		-0.0010	-0.0033				
a_{13}									
a_{14}									
a_{15}									
sum	344.8019	-2.4299	0.9876	393.1320	1.9117	0.9436	400.9347	2.6023	0.9378
n	13	13	8	13	13	8	12	12	7

	April			May			June		
	λ °	β °	HP °	λ °	β °	HP °	λ °	β °	HP °
a_0	235.2509	1.6505	0.9402	269.5307	2.2028	0.9328	316.2905	1.8159	0.9317
a_1	211.1416	-0.5660	0.0024	211.7679	-0.2102	-0.0013	211.8658	0.3268	-0.0076
a_2	3.3937	3.0516	-0.0225	0.9870	4.1772	-0.0401	-2.9934	3.2509	-0.0428
a_3	-2.7567	3.0664	-0.0278	-4.6160	0.8004	-0.0093	-4.7207	-2.4626	0.0213
a_4	-1.7453	-1.6400	0.0149	-0.4216	-2.3847	0.0236	1.8502	-1.8341	0.0216
a_5	0.8816	-0.9003	0.0040	1.4275	-0.2603	-0.0012	1.0940	0.7984	-0.0113
a_6	0.1522	0.2128	-0.0040	-0.1550	0.5025	-0.0062	-0.7750	0.3335	-0.0030
a_7	-0.1902	0.2012		-0.3495	0.0709	0.0019	-0.0989	-0.2178	0.0048
a_8	-0.0012	-0.0152		0.0850	-0.1152	0.0017	0.2674	-0.0379	
a_9	0.0443	-0.0451		0.0941	-0.0119		-0.0140	0.0684	
a_{10}	0.0042	0.0048		-0.0233	0.0326		-0.0729	-0.0055	
a_{11}	-0.0105	0.0093		-0.0253	-0.0004		0.0158	-0.0205	
a_{12}	-0.0021	-0.0021		0.0064	-0.0097		0.0174	0.0065	
a_{13}	0.0023	-0.0023		0.0067	0.0010		-0.0080	0.0053	
a_{14}				-0.0020	0.0029		-0.0035	-0.0031	
a_{15}				-0.0018			0.0032		
sum	446.1648	5.0256	0.9072	478.3108	4.7979	0.9019	522.7179	2.0242	0.9147
n	14	14	7	16	15	9	16	15	8

Evaluate the recurrence relation :

$b_i = 2(2x-1)b_{i+1} - b_{i+2} + a_i$ for $i = n-1, n-2, \ldots 2, 1, 0$

where $b_{n+1} = b_n = 0$ and $x = (d + GMT^h/24)/32$

Then $\lambda^\circ, \beta^\circ, HP^\circ = (b_0 - b_2 + a_0)/2$

	July			August			September		
	λ °	β °	HP °	λ °	β °	HP °	λ °	β °	HP °
a_0	350.6240	0.7905	0.9370	38.6100	-0.9323	0.9506	88.0940	-1.9258	0.9640
a_1	211.6861	0.5225	-0.0106	211.0553	0.3527	-0.0115	210.6175	-0.0599	-0.0079
a_2	-5.2143	0.9915	-0.0315	-6.3089	-2.6485	-0.0013	-4.0949	-4.6380	0.0276
a_3	-3.3196	-3.8891	0.0381	0.3228	-3.3790	0.0458	3.6253	-0.2858	0.0290
a_4	3.0040	-0.4801	0.0129	3.2165	1.5284	-0.0065	1.7514	2.4380	-0.0209
a_5	0.4127	1.1212	-0.0154	-0.7355	0.6844	-0.0124	-1.2580	-0.0936	-0.0032
a_6	-0.8999	-0.0739	0.0011	-0.4592	-0.3789	0.0069	0.0101	-0.2113	0.0068
a_7	0.1879	-0.2121	0.0051	0.3979	0.0496	0.0014	0.2353	0.1001	-0.0018
a_8	0.2281	0.0894	-0.0017	-0.0167	0.0741	-0.0026	-0.0913	-0.0580	
a_9	-0.1079	0.0326		-0.0947	-0.0550		-0.0102	-0.0156	
a_{10}	-0.0354	-0.0392		0.0365	-0.0007		0.0185	0.0240	
a_{11}	0.0386	0.0013		0.0069	0.0174		-0.0086	-0.0048	
a_{12}	-0.0016	0.0125		-0.0129	-0.0064		0.0002	-0.0035	
a_{13}	-0.0108	-0.0039		0.0039	-0.0026		0.0023	0.0025	
a_{14}	0.0038	-0.0029		0.0020	0.0029				
a_{15}	0.0022	0.0021							
sum	556.5979	-1.1376	0.9350	246.0239	-4.6939	0.9704	298.8916	-4.7317	0.9936
n	16	16	9	15	15	9	14	14	8

	October			November			December		
	λ °	β °	HP °	λ °	β °	HP °	λ °	β °	HP °
a_0	124.8771	-1.6242	0.9673	175.1297	-0.0248	0.9616	211.1340	1.2988	0.9514
a_1	210.6289	-0.2802	-0.0030	210.8379	-0.4736	0.0025	211.2797	-0.4170	0.0069
a_2	-0.7875	-3.9268	0.0340	3.1929	-0.5448	0.0213	5.0216	2.2539	-0.0010
a_3	4.2858	2.3743	0.0044	2.6890	4.2104	-0.0268	0.2254	3.4536	-0.0398
a_4	0.1045	2.0284	-0.0235	-2.0960	0.3077	-0.0155	-3.0772	-1.1876	-0.0040
a_5	-1.3487	-0.5158	0.0021	-1.0127	-0.9135	0.0102	-0.4346	-0.9420	0.0146
a_6	0.0978	-0.1342	0.0057	0.4952	-0.1507	0.0045	0.8228	0.0138	0.0038
a_7	0.2085	-0.0140		0.2702	-0.0047	-0.0020	0.2546	0.1477	-0.0034
a_8	-0.0257	-0.0545		-0.0528	0.0626		-0.1576	0.0857	
a_9	-0.0073	0.0277		-0.0553	0.0357		-0.0951	-0.0172	
a_{10}	0.0004	0.0141		-0.0143	-0.0143		0.0064	-0.0340	
a_{11}	-0.0060	-0.0069		0.0057	-0.0107		0.0272	-0.0033	
a_{12}				0.0083	0.0004		0.0087	0.0085	
a_{13}				0.0010	0.0021		-0.0055	0.0035	
a_{14}							-0.0043	-0.0012	
a_{15}									
sum	338.0278	-2.1121	0.9870	389.3988	2.4818	0.9558	425.0061	4.6632	0.9285
n	12	12	7	14	14	8	15	15	8

Calculate $X = \cos\beta\cos\lambda$ $Y = \cos\epsilon\cos\beta\sin\lambda - \sin\epsilon\sin\beta$

and $Z = \sin\epsilon\cos\beta\sin\lambda + \cos\epsilon\sin\beta$ where $\epsilon = 23°.44$

Then $RA° = \tan^{-1}(Y/X)$ and $DEC° = \sin^{-1}(Z)$

Add 180° to RA if $X < 0$. Add 360° to RA if $X > 0$ and $Y < 0$.

Table 3: Moon Monthly Chebyshev Coefficients, 1988

	January			February			March		
	λ°	β°	HP°	λ°	β°	HP°	λ°	β°	HP°
a_0	259.2543	2.1736	0.9377	306.2696	1.6304	0.9305	327.6437	0.9749	0.9307
a_1	211.9396	-0.0447	0.0075	212.2834	0.4701	0.0038	212.0575	0.6233	0.0031
a_2	4.5673	4.1883	-0.0313	1.6975	3.1515	-0.0472	0.4173	1.8980	-0.0457
a_3	-3.3574	0.4385	-0.0350	-5.4768	-2.8043	-0.0118	-5.3473	-3.7709	-0.0013
a_4	-2.8678	-2.3877	0.0157	-1.0885	-1.8521	0.0281	-0.3461	-1.0857	0.0270
a_5	0.8823	-0.2902	0.0157	1.7411	0.7744	0.0074	1.5886	1.0922	0.0044
a_6	0.9316	0.4473	-0.0025	0.4063	0.4576	-0.0089	0.2844	0.2525	-0.0080
a_7	-0.1347	0.1685	-0.0057	-0.4475	-0.1421	-0.0036	-0.3505	-0.2352	-0.0028
a_8	-0.2770	-0.0694		-0.1335	-0.1352	0.0027	-0.1093	-0.0758	0.0024
a_9	-0.0042	-0.0723		0.1151	0.0288		0.0814	0.0592	
a_{10}	0.0732	0.0072		0.0413	0.0465		0.0318	0.0290	
a_{11}	0.0158	0.0251		-0.0290	-0.0052		-0.0187	-0.0153	
a_{12}	-0.0180	0.0026		-0.0131	-0.0157		-0.0092	-0.0106	
a_{13}	-0.0084	-0.0074		0.0075	0.0004		0.0042	0.0039	
a_{14}	0.0039	-0.0024		0.0042	0.0051		0.0027	0.0036	
a_{15}	0.0033	0.0019		-0.0020					
sum	471.0038	4.5789	0.9021	515.3756	1.6102	0.9010	535.9305	-0.2569	0.9098
n	16	16	8	16	15	9	15	15	9

	April			May			June		
	λ°	β°	HP°	λ°	β°	HP°	λ°	β°	HP°
a_0	14.5089	-0.6872	0.9382	49.3209	-1.6472	0.9491	98.4764	-1.8301	0.9652
a_1	211.6046	0.5937	-0.0017	210.9415	0.2687	-0.0027	210.2180	-0.2255	-0.0002
a_2	-3.0809	-1.6148	-0.0288	-4.2992	-3.6624	-0.0039	-3.1044	-4.0572	0.0318
a_3	-3.5375	-3.8510	0.0251	-0.7064	-2.1287	0.0350	3.2848	1.4915	0.0278
a_4	1.7152	0.9911	0.0191	2.3919	2.0819	0.0061	2.0185	2.1601	-0.0118
a_5	1.1793	1.0066	-0.0060	0.5333	0.3866	-0.0082	-0.3026	-0.4205	-0.0077
a_6	-0.1957	-0.2987	-0.0065	-0.2638	-0.4002	-0.0034	-0.3898	-0.1521	-0.0017
a_7	-0.2878	-0.1583	0.0010	-0.2131	0.0451		-0.1656	0.0952	
a_8	0.0096	0.0919	0.0020	0.0131	0.0766		0.0600	-0.0376	
a_9	0.0705	0.0357		0.0546	-0.0218		0.0269	-0.0243	
a_{10}	-0.0038	-0.0250		-0.0121	-0.0172		-0.0144	0.0044	
a_{11}	-0.0180	-0.0103		-0.0122	0.0037		0.0052	0.0046	
a_{12}	0.0016	0.0069		0.0056	0.0041		0.0039		
a_{13}	0.0047	0.0029		0.0026					
a_{14}		-0.0021							
a_{15}									
sum	221.9707	-3.9186	0.9424	257.7567	-5.0108	0.9720	310.1169	-2.9915	1.0034
n	14	15	9	14	13	7	13	12	7

Evaluate the recurrence relation :

$b_i = 2(2x-1)b_{i+1} - b_{i+2} + a_i$ for $i=n-1, n-2, \ldots 2, 1, 0$

where $b_{n+1} = b_n = 0$ and $x = (d + \mathrm{GMT}^h/24)/32$

Then $\lambda°, \beta°, HP° = (b_0 - b_2 + a_0)/2$

	July			August			September		
	λ °	β °	HP °	λ °	β °	HP °	λ °	β °	HP °
a_0	135.0175	-0.9781	0.9736	185.5334	0.8393	0.9719	235.2366	2.0019	0.9580
a_1	209.6874	-0.4905	0.0027	209.5998	-0.5850	0.0078	210.0412	-0.3546	0.0118
a_2	-0.6557	-2.2291	0.0508	3.7082	1.7214	0.0478	6.4562	4.3681	0.0184
a_3	5.2791	3.7263	0.0090	5.1561	4.1257	-0.0255	2.2748	1.4661	-0.0463
a_4	0.8657	1.1895	-0.0206	-1.1942	-0.7610	-0.0226	-2.7178	-2.2352	-0.0126
a_5	-0.7522	-0.7129	-0.0054	-1.0745	-0.8351	-0.0025	-0.8228	-0.5526	0.0038
a_6	-0.3704	-0.0360	-0.0020	-0.2947	-0.0761	0.0011	-0.0271	0.1558	0.0055
a_7	-0.1310	-0.0223		0.0289	-0.0278	0.0036	0.1960	0.1507	0.0031
a_8	0.0740	-0.0483		0.1249	0.0403		0.1279	0.0596	
a_9	0.0272	0.0047		0.0267	0.0306		-0.0072	-0.0076	
a_{10}	-0.0011	0.0083		-0.0049	0.0088		-0.0250	-0.0190	
a_{11}	0.0050	0.0050		-0.0064	-0.0032		-0.0102	-0.0103	
a_{12}	-0.0013			-0.0042	-0.0049		0.0003	0.0009	
a_{13}							0.0025	0.0034	
a_{14}									
a_{15}									
sum	349.0442	0.4166	1.0081	401.5991	4.4730	0.9816	450.7254	5.0272	0.9417
n	13	12	7	13	13	8	14	14	8

	October			November			December		
	λ °	β °	HP °	λ °	β °	HP °	λ °	β °	HP °
a_0	270.4526	1.8965	0.9471	317.6641	0.7022	0.9348	351.4894	-0.5154	0.9318
a_1	210.3987	0.0037	0.0114	211.0016	0.4654	0.0070	211.1397	0.5302	0.0002
a_2	6.3023	4.2669	-0.0051	3.5533	1.7155	-0.0325	0.1983	-1.0015	-0.0390
a_3	-0.3451	-1.1594	-0.0449	-3.5104	-3.6569	-0.0250	-4.2937	-3.8742	-0.0017
a_4	-2.8018	-2.3433	-0.0010	-1.7004	-1.0292	0.0136	-0.0955	0.5504	0.0177
a_5	-0.2966	0.0676	0.0067	0.3912	0.9677	0.0062	0.6431	1.1094	
a_6	0.0892	0.3992	0.0049	0.2036	0.3026	0.0018	0.0198	-0.1070	
a_7	0.2072	0.1252	0.0014	0.2022	-0.1321		0.1436	-0.2077	
a_8	0.1069	-0.0294		0.0163	-0.0799		-0.0069	0.0174	
a_9	-0.0276	-0.0369		-0.0778	0.0052		-0.0733	0.0318	
a_{10}	-0.0336	-0.0127		-0.0212	0.0135		0.0021	-0.0017	
a_{11}	-0.0071	0.0020		0.0160	0.0060		0.0203	-0.0007	
a_{12}	0.0048	0.0059		0.0084	-0.0001		-0.0005		
a_{13}	0.0035	0.0021		-0.0017	-0.0026		-0.0043		
a_{14}									
a_{15}									
sum	484.0534	3.1874	0.9205	527.7452	-0.7227	0.9059	559.1821	-3.4690	0.9090
n	14	14	8	14	14	7	14	12	5

Calculate $X = \cos\beta \cos\lambda$ $Y = \cos\epsilon \cos\beta \sin\lambda - \sin\epsilon \sin\beta$

and $Z = \sin\epsilon \cos\beta \sin\lambda + \cos\epsilon \sin\beta$ where $\epsilon = 23°.44$

Then $RA° = \tan^{-1}(Y/X)$ and $DEC° = \sin^{-1}(Z)$

Add 180° to RA if $X < 0$. Add 360° to RA if $X > 0$ and $Y < 0$.

Table 3: Moon Monthly Chebyshev Coefficients, 1989

	January			February			March		
	λ °	β °	HP °	λ °	β °	HP °	λ °	β °	HP °
a_0	38.0110	-1.8372	0.9364	85.4177	-2.0591	0.9500	94.1642	-1.9638	0.9495
a_1	210.8840	0.1677	-0.0077	210.1906	-0.3726	-0.0113	210.5471	-0.3754	-0.0132
a_2	-4.1043	-3.9007	-0.0286	-6.3148	-4.2780	0.0017	-6.7168	-3.9489	-0.0010
a_3	-3.0114	-1.8638	0.0289	0.3712	1.5665	0.0456	0.3210	2.1509	0.0481
a_4	1.9027	2.1986	0.0116	2.7700	2.2195	-0.0037	3.1262	2.0282	-0.0058
a_5	0.3588	0.3791	-0.0060	-0.4019	-0.5942	-0.0054	-0.6674	-0.7415	-0.0088
a_6	-0.1503	-0.4725	0.0023	-0.0910	-0.2186	0.0047	-0.3246	-0.1568	0.0071
a_7	0.1392	0.0158		0.1897	0.1739	-0.0021	0.3517	0.1861	-0.0009
a_8	-0.0446	0.0827		-0.0898	-0.0338		-0.0413	-0.0623	-0.0023
a_9	-0.0439	-0.0210		-0.0235	-0.0207		-0.0756	-0.0205	
a_{10}	0.0246	-0.0053		0.0252	0.0174		0.0324	0.0288	
a_{11}	0.0049	0.0059		-0.0053	-0.0062		0.0042	-0.0081	
a_{12}	-0.0071	-0.0030		-0.0031	-0.0026		-0.0096	-0.0057	
a_{13}				0.0024	0.0032		0.0029	0.0047	
a_{14}							0.0013		
a_{15}									
sum	243.9636	-5.2537	0.9369	292.0374	-3.6053	0.9795	300.7157	-2.8843	0.9727
n	13	13	7	14	14	8	15	14	9

	April			May			June		
	λ °	β °	HP °	λ °	β °	HP °	λ °	β °	HP °
a_0	143.1877	-0.5731	0.9650	179.7210	0.7861	0.9705	230.2915	1.8678	0.9680
a_1	210.0069	-0.5178	-0.0105	210.0966	-0.3602	-0.0068	210.3533	-0.0226	-0.0016
a_2	-5.3216	-0.8234	0.0322	-2.6355	2.1107	0.0427	1.7220	4.3823	0.0367
a_3	3.7677	4.1932	0.0374	4.8752	3.6791	0.0163	4.0784	0.5670	-0.0174
a_4	2.0422	0.2707	-0.0199	0.6183	-1.1903	-0.0236	-1.4917	-2.3255	-0.0174
a_5	-1.2459	-0.9064	-0.0009	-1.2324	-0.6845	0.0034	-0.6644	-0.1886	0.0082
a_6	0.1941	0.1558	0.0052	0.4227	0.1452	0.0030	0.4627	0.1704	
a_7	0.2582	-0.0081	-0.0037	0.1081	-0.0632	-0.0032	-0.0892	0.0673	
a_8	-0.1628	-0.0564		-0.1557	0.0249		-0.0821	0.0429	
a_9	-0.0065	0.0396		0.0185	0.0248		0.0189	-0.0251	
a_{10}	0.0276	-0.0055		0.0063	-0.0177		0.0054	-0.0071	
a_{11}	-0.0148	-0.0094		-0.0077	0.0020		0.0065	0.0056	
a_{12}	0.0017	0.0063		0.0066	0.0033				
a_{13}	0.0036								
a_{14}	-0.0022								
a_{15}									
sum	352.7359	1.7655	1.0048	391.8420	4.4602	1.0023	444.6113	4.5344	0.9765
n	15	13	8	13	13	8	12	12	6

Evaluate the recurrence relation :

$b_i = 2(2x-1)b_{i+1} - b_{i+2} + a_i$ for $i=n-1, n-2, \ldots 2, 1, 0$

where $b_{n+1} = b_n = 0$ and $x = (d + \mathrm{GMT}^h/24)/32$

Then $\lambda°, \beta°, \mathrm{HP}° = (b_0 - b_2 + a_0)/2$

	July			August			September		
	λ°	β°	HP°	λ°	β°	HP°	λ°	β°	HP°
a_0	266.6967	1.6331	0.9579	315.3069	0.1950	0.9424	2.5629	-1.4358	0.9319
a_1	210.8767	0.3034	0.0019	211.5719	0.6456	0.0033	212.1776	0.5633	0.0002
a_2	3.9010	3.8863	0.0140	4.0195	0.8590	-0.0203	1.2984	-2.5622	-0.0440
a_3	1.5293	-2.1500	-0.0329	-2.5091	-4.1684	-0.0328	-5.1965	-3.3059	-0.0117
a_4	-2.5015	-2.1369	-0.0060	-2.4486	-0.5534	0.0136	-0.7857	1.4674	0.0273
a_5	-0.1805	0.3515	0.0114	0.8256	1.0659	0.0110	1.7106	0.9852	0.0025
a_6	0.4973	0.3292	-0.0002	0.5395	0.2479	-0.0047	0.1678	-0.2861	-0.0093
a_7	-0.0546	0.0854	-0.0024	-0.1716	-0.1489	-0.0031	-0.4541	-0.2368	-0.0003
a_8	-0.0698	-0.0400		-0.1205	-0.1000		-0.0528	0.0616	0.0030
a_9	0.0036	-0.0411		0.0225	0.0251		0.1191	0.0705	
a_{10}	0.0123	0.0069		0.0307	0.0313		0.0204	-0.0190	
a_{11}	0.0048	0.0091		-0.0002	-0.0043		-0.0307	-0.0223	
a_{12}	-0.0030	-0.0003		-0.0074	-0.0090		-0.0073	0.0059	
a_{13}				-0.0010	0.0001		0.0082	0.0071	
a_{14}					0.0026		0.0025	-0.0017	
a_{15}							-0.0023	-0.0023	
sum	480.7123	2.2366	0.9437	527.0582	-1.9115	0.9094	211.5381	-4.7111	0.8996
n	13	13	8	14	15	8	16	16	9

	October			November			December		
	λ°	β°	HP°	λ°	β°	HP°	λ°	β°	HP°
a_0	36.6801	-2.1330	0.9301	83.6789	-1.9239	0.9365	118.5401	-0.9336	0.9463
a_1	212.3802	0.2139	-0.0027	212.0309	-0.3103	-0.0067	211.5097	-0.5376	-0.0057
a_2	-1.1397	-3.9515	-0.0484	-4.1850	-3.4381	-0.0340	-4.7830	-1.3358	-0.0122
a_3	-5.5972	-1.1888	0.0079	-3.7683	2.1869	0.0326	-1.2946	3.9008	0.0386
a_4	0.7437	2.3219	0.0293	2.6829	1.9498	0.0182	3.0554	0.6991	0.0053
a_5	1.8846	0.3096	-0.0051	1.0499	-0.7371	-0.0147	0.2067	-1.0516	-0.0154
a_6	-0.2962	-0.5417	-0.0098	-0.9323	-0.3124	-0.0041	-0.8892	0.0507	-0.0003
a_7	-0.5417	-0.0436	0.0025	-0.1848	0.2256	0.0055	0.0697	0.1716	0.0046
a_8	0.1029	0.1474	0.0031	0.2965	0.0232	0.0004	0.2157	-0.0877	
a_9	0.1532	0.0029		0.0076	-0.0772		-0.0627	-0.0288	
a_{10}	-0.0339	-0.0484		-0.0827	0.0076		-0.0367	0.0351	
a_{11}	-0.0426	0.0020		0.0142	0.0242		0.0276	0.0018	
a_{12}	0.0116	0.0160		0.0215	-0.0067		0.0020	-0.0106	
a_{13}	0.0124	-0.0015		-0.0087	-0.0065		-0.0088	0.0021	
a_{14}	-0.0041	-0.0052		-0.0050	0.0034		0.0017	0.0026	
a_{15}	-0.0038	0.0008		0.0037	0.0014		0.0021		
sum	244.3095	-4.8992	0.9069	290.6193	-2.3901	0.9337	326.5557	0.8781	0.9612
n	16	16	9	16	16	9	16	15	8

Calculate $X = \cos\beta\cos\lambda$ $Y = \cos\epsilon\cos\beta\sin\lambda - \sin\epsilon\sin\beta$

and $Z = \sin\epsilon\cos\beta\sin\lambda + \cos\epsilon\sin\beta$ where $\epsilon = 23°.44$

Then $RA° = \tan^{-1}(Y/X)$ and $DEC° = \sin^{-1}(Z)$

Add 180° to RA if $X < 0$. Add 360° to RA if $X > 0$ and $Y < 0$.

Table 3: Moon Monthly Chebyshev Coefficients, 1990

	January			February			March		
	λ°	β°	HP°	λ°	β°	HP°	λ°	β°	HP°
a_0	167.5018	0.8564	0.9606	217.6824	1.8950	0.9675	227.1063	1.8604	0.9708
a_1	210.8588	-0.4502	-0.0024	210.5821	-0.1371	0.0031	210.0852	-0.1122	0.0062
a_2	-3.4199	2.4203	0.0194	0.5125	4.6120	0.0348	1.7714	4.5663	0.0436
a_3	2.4238	3.7087	0.0286	4.3630	0.6830	-0.0014	5.0078	-0.0711	-0.0093
a_4	2.2540	-1.3130	-0.0136	0.1808	-2.4042	-0.0234	-0.1603	-2.3771	-0.0233
a_5	-0.8359	-0.7497	-0.0112	-1.2640	-0.1535	-0.0038	-1.0639	-0.0417	-0.0049
a_6	-0.5004	0.2489	0.0036	-0.2469	0.1707	0.0052	-0.4052	0.1463	0.0024
a_7	0.2001	-0.0461	0.0024	0.1459	-0.0111		-0.0110	0.0172	0.0031
a_8	0.0490	-0.0472		0.0606	0.0613		0.1162	0.0594	
a_9	-0.0430	0.0446		0.0037	0.0182		0.0306	0.0147	
a_{10}	0.0107	0.0073		0.0004	-0.0167		0.0003	-0.0102	
a_{11}	0.0060	-0.0114		-0.0040	-0.0057		-0.0014	-0.0079	
a_{12}	-0.0058	0.0011		-0.0029			-0.0044		
a_{13}		0.0018							
a_{14}									
a_{15}									
sum	378.4992	4.6715	0.9874	432.0136	4.7119	0.9820	442.4716	4.0441	0.9886
n	13	14	8	13	12	7	13	12	8

	April			May			June		
	λ°	β°	HP°	λ°	β°	HP°	λ°	β°	HP°
a_0	277.3158	0.9089	0.9627	313.1204	-0.3967	0.9528	1.0208	-1.8689	0.9382
a_1	210.2294	0.2880	0.0110	210.4448	0.4975	0.0128	211.1583	0.3737	0.0113
a_2	5.6123	2.6250	0.0266	6.7182	-0.1783	0.0058	5.3838	-3.4592	-0.0263
a_3	3.2895	-3.3293	-0.0391	0.9608	-4.0962	-0.0481	-2.6310	-2.5149	-0.0387
a_4	-2.2295	-1.4948	-0.0188	-3.0307	-0.0710	-0.0088	-2.8007	1.8376	0.0089
a_5	-1.2300	0.5949	0.0027	-0.8907	0.9764	0.0076	0.1416	0.8390	0.0117
a_6	-0.1895	0.2773	0.0064	0.1735	0.2295	0.0070	0.6379	-0.2240	0.0037
a_7	0.2871	0.0864	0.0031	0.4180	-0.0507	0.0015	0.2530	-0.2193	-0.0024
a_8	0.1838	-0.0105		0.1100	-0.0882	-0.0020	-0.1206	-0.0238	-0.0023
a_9	-0.0081	-0.0405		-0.0761	-0.0338		-0.1013	0.0464	
a_{10}	-0.0338	-0.0190		-0.0498	0.0082		0.0071	0.0252	
a_{11}	-0.0175	0.0026		-0.0045	0.0158		0.0268	-0.0039	
a_{12}	-0.0019	0.0070		0.0108	0.0047		0.0061	-0.0097	
a_{13}	0.0043	0.0024		0.0066	-0.0028		-0.0052	-0.0021	
a_{14}	0.0029			0.0001	-0.0025		-0.0033	0.0023	
a_{15}				-0.0019					
sum	493.2148	-0.1016	0.9546	527.9095	-3.1881	0.9286	212.9733	-5.2016	0.9041
n	15	14	8	16	15	9	15	15	9

Evaluate the recurrence relation :

$b_i = 2(2x-1)b_{i+1} - b_{i+2} + a_i$ for $i=n-1, n-2, \ldots 2, 1, 0$

where $b_{n+1} = b_n = 0$ and $x = (d + \mathrm{GMT}^h/24)/32$

Then λ°, β°, $HP^\circ = (b_0 - b_2 + a_0)/2$

	July			August			September		
	λ °	β °	HP °	λ °	β °	HP °	λ °	β °	HP °
a_0	35.2683	-2.2532	0.9328	81.9218	-1.6465	0.9336	128.9951	-0.1199	0.9449
a_1	211.3697	0.0310	0.0066	211.2688	-0.4889	-0.0009	210.5574	-0.6110	-0.0059
a_2	2.8228	-4.4214	-0.0381	-1.4803	-3.1557	-0.0362	-4.7883	0.1943	-0.0105
a_3	-4.0745	-0.1765	-0.0197	-4.0701	2.9425	0.0122	-1.3766	4.1326	0.0364
a_4	-1.5423	2.4701	0.0173	0.6473	1.7251	0.0185	2.2114	-0.2474	0.0067
a_5	0.7522	0.1568	0.0079	0.8715	-0.8670	0.0001	0.2325	-1.0365	-0.0045
a_6	0.4495	-0.5010	0.0003	0.0435	-0.3150	-0.0018	-0.0306	0.2038	
a_7	0.0280	-0.0950	-0.0026	-0.0367	0.1819		0.0592	0.1236	
a_8	-0.1193	0.0856		-0.0447	0.0539		-0.0906	-0.0685	
a_9	-0.0342	0.0389		-0.0095	-0.0325		-0.0105	-0.0085	
a_{10}	0.0247	-0.0082		0.0131	-0.0115		0.0271	0.0077	
a_{11}	0.0119	-0.0125		0.0035	0.0029		-0.0043	-0.0008	
a_{12}	-0.0038	-0.0021		-0.0031	0.0029		-0.0054		
a_{13}	-0.0034	0.0032							
a_{14}									
a_{15}									
sum	244.9496	-4.6843	0.9045	289.1251	-1.6079	0.9255	335.7764	2.5694	0.9671
n	14	14	8	13	13	7	13	12	6

	October			November			December		
	λ °	β °	HP °	λ °	β °	HP °	λ °	β °	HP °
a_0	163.9205	1.0601	0.9566	213.3009	1.8353	0.9725	249.7019	1.4163	0.9771
a_1	209.8788	-0.3482	-0.0078	209.2108	0.1540	-0.0048	209.0367	0.4750	-0.0032
a_2	-5.7663	2.7944	0.0161	-3.7046	4.4370	0.0510	-1.1533	3.4672	0.0609
a_3	1.5573	3.1167	0.0428	5.2006	-0.2392	0.0279	6.1732	-2.9035	0.0074
a_4	2.4823	-1.6130	-0.0060	1.4915	-2.2973	-0.0189	0.2579	-1.7514	-0.0223
a_5	-0.3225	-0.5651	-0.0034	-0.5617	0.1763	0.0001	-0.6706	0.5388	0.0023
a_6	0.1538	0.3336	0.0012	0.1872	0.1252	-0.0020	0.1107	0.0477	-0.0036
a_7	0.0333	-0.0545	-0.0026	-0.2229	-0.0775	-0.0026	-0.2485	0.0328	-0.0015
a_8	-0.1555	-0.0431		-0.0879	0.0432		-0.0237	0.0378	
a_9	0.0258	0.0218		0.0557	0.0010		0.0453	-0.0124	
a_{10}	0.0227	-0.0054		-0.0080	0.0011		-0.0035	0.0070	
a_{11}	-0.0149	0.0012		-0.0010	0.0055		0.0050	-0.0021	
a_{12}	0.0011	0.0031		0.0047	-0.0035			-0.0036	
a_{13}	0.0024								
a_{14}									
a_{15}									
sum	371.8188	4.7016	0.9969	424.8653	4.1611	1.0232	463.2311	1.3496	1.0171
n	14	13	8	13	13	8	12	13	8

Calculate $X = \cos\beta \cos\lambda$ $\quad Y = \cos\epsilon \cos\beta \sin\lambda - \sin\epsilon \sin\beta$

and $Z = \sin\epsilon \cos\beta \sin\lambda + \cos\epsilon \sin\beta$ where $\epsilon = 23°.44$

Then $RA° = \tan^{-1}(Y/X)$ and $DEC° = \sin^{-1}(Z)$

Add 180° to RA if $X < 0$. Add 360° to RA if $X > 0$ and $Y < 0$.

TABLE 4: PHASES OF THE MOON AND SEASONS

New Moon	First Quarter	Full Moon	Last Quarter	Seasons
		1986		1986
d h m	d h m	d h m	d h m	
			Jan. 3 19 47	
Jan. 10 12 22	Jan. 17 22 13	Jan. 26 00 31	Feb. 2 04 41	Equinoxes
Feb. 9 00 55	Feb. 16 19 55	Feb. 24 15 02	Mar. 3 12 17	
Mar. 10 14 52	Mar. 18 16 39	Mar. 26 03 02	Apr. 1 19 30	d h m
Apr. 9 06 08	Apr. 17 10 35	Apr. 24 12 46	May 1 03 22	Mar. 20 22 03
May 8 22 10	May 17 01 00	May 23 20 45	May 30 12 55	Sept. 23 07 59
June 7 14 00	June 15 12 00	June 22 03 42	June 29 00 53	
July 7 04 55	July 14 20 10	July 21 10 40	July 28 15 34	
Aug. 5 18 36	Aug. 13 02 21	Aug. 19 18 54	Aug. 27 08 38	Solstices
Sept. 4 07 10	Sept.11 07 41	Sept.18 05 34	Sept.26 03 17	
Oct. 3 18 55	Oct. 10 13 28	Oct. 17 19 22	Oct. 25 22 26	d h m
Nov. 2 06 02	Nov. 8 21 11	Nov. 16 12 12	Nov. 24 16 50	June 21 16 30
Dec. 1 16 43	Dec. 8 08 01	Dec. 16 07 04	Dec. 24 09 17	Dec. 22 04 02
Dec. 31 03 10				
		1987		1987
	Jan. 6 22 34	Jan. 15 02 30	Jan. 22 22 45	Equinoxes
Jan. 29 13 44	Feb. 5 16 21	Feb. 13 20 58	Feb. 21 08 56	
Feb. 28 00 51	Mar. 7 11 58	Mar. 15 13 13	Mar. 22 16 22	d h m
Mar. 29 12 46	Apr. 6 07 48	Apr. 14 02 31	Apr. 20 22 15	Mar. 21 03 52
Apr. 28 01 34	May 6 02 26	May 13 12 50	May 20 04 02	Sept. 23 13 45
May 27 15 13	June 4 18 53	June 11 20 49	June 18 11 02	
June 26 05 37	July 4 08 34	July 11 03 33	July 17 20 17	
July 25 20 37	Aug. 2 19 24	Aug. 9 10 17	Aug. 16 08 25	Solstices
Aug. 24 11 59	Sept. 1 03 48	Sept. 7 18 13	Sept.14 23 44	
Sept.23 03 08	Sept.30 10 39	Oct. 7 04 12	Oct. 14 18 06	d h m
Oct. 22 17 28	Oct. 29 17 10	Nov. 5 16 46	Nov. 13 14 38	June 21 22 11
Nov. 21 06 33	Nov. 28 00 37	Dec. 5 08 01	Dec. 13 11 41	Dec. 22 09 46
Dec. 20 18 25	Dec. 27 10 01			
		1988		1988
		Jan. 4 01 40	Jan. 12 07 04	
Jan. 19 05 26	Jan. 25 21 53	Feb. 2 20 51	Feb. 10 23 01	Equinoxes
Feb. 17 15 54	Feb. 24 12 15	Mar. 3 16 01	Mar. 11 10 56	
Mar. 18 02 02	Mar. 25 04 41	Apr. 2 09 21	Apr. 9 19 21	d h m
Apr. 16 12 00	Apr. 23 22 32	May 1 23 41	May 9 01 23	Mar. 20 09 39
May 15 22 11	May 23 16 49	May 31 10 53	June 7 06 21	Sept. 22 19 29
June 14 09 14	June 22 10 23	June 29 19 46	July 6 11 36	
July 13 21 53	July 22 02 14	July 29 03 25	Aug. 4 18 22	
Aug. 12 12 31	Aug. 20 15 51	Aug. 27 10 56	Sept. 3 03 50	Solstices
Sept.11 04 49	Sept.19 03 18	Sept.25 19 07	Oct. 2 16 58	
Oct. 10 21 49	Oct. 18 13 01	Oct. 25 04 35	Nov. 1 10 11	d h m
Nov. 9 14 20	Nov. 16 21 35	Nov. 23 15 53	Dec. 1 06 49	June 21 03 57
Dec. 9 05 36	Dec. 16 05 40	Dec. 23 05 29	Dec. 31 04 57	Dec. 21 15 28
		1989		1989
Jan. 7 19 22	Jan. 14 13 58	Jan. 21 21 33	Jan. 30 02 02	Equinoxes
Feb. 6 07 37	Feb. 12 23 15	Feb. 20 15 32	Feb. 28 20 08	
Mar. 7 18 19	Mar. 14 10 11	Mar. 22 09 58	Mar. 30 10 21	d h m
Apr. 6 03 33	Apr. 12 23 13	Apr. 21 03 13	Apr. 28 20 46	Mar. 20 15 28
May 5 11 46	May 12 14 19	May 20 18 16	May 28 04 01	Sept. 23 01 20
June 3 19 53	June 11 06 59	June 19 06 57	June 26 09 09	
July 3 04 59	July 11 00 19	July 18 17 42	July 25 13 31	
Aug. 1 16 06	Aug. 9 17 28	Aug. 17 03 07	Aug. 23 18 40	Solstices
Aug. 31 05 44	Sept. 8 09 49	Sept.15 11 51	Sept.22 02 10	
Sept.29 21 47	Oct. 8 00 52	Oct. 14 20 32	Oct. 21 13 19	d h m
Oct. 29 15 27	Nov. 6 14 11	Nov. 13 05 51	Nov. 20 04 44	June 21 09 53
Nov. 28 09 41	Dec. 6 01 26	Dec. 12 16 30	Dec. 19 23 54	Dec. 21 21 22
Dec. 28 03 20				

TABLE 4: PHASES OF THE MOON AND SEASONS

1990

New Moon	First Quarter	Full Moon	Last Quarter
d h m	d h m	d h m	d h m
	Jan. 4 10 40	Jan. 11 04 57	Jan. 18 21 17
Jan. 26 19 20	Feb. 2 18 32	Feb. 9 19 16	Feb. 17 18 48
Feb. 25 08 54	Mar. 4 02 05	Mar. 11 10 58	Mar. 19 14 30
Mar. 26 19 48	Apr. 2 10 24	Apr. 10 03 18	Apr. 18 07 02
Apr. 25 04 27	May 1 20 18	May 9 19 31	May 17 19 45
May 24 11 47	May 31 08 11	June 8 11 01	June 16 04 48
June 22 18 55	June 29 22 07	July 8 01 23	July 15 11 04
July 22 02 54	July 29 14 01	Aug. 6 14 19	Aug. 13 15 54
Aug. 20 12 39	Aug. 28 07 34	Sept. 5 01 46	Sept.11 20 53
Sept.19 00 46	Sept.27 02 06	Oct. 4 12 02	Oct. 11 03 31
Oct. 18 15 37	Oct. 26 20 26	Nov. 2 21 48	Nov. 9 13 02
Nov. 17 09 05	Nov. 25 13 11	Dec. 2 07 50	Dec. 9 02 04
Dec. 17 04 22	Dec. 25 03 16	Dec. 31 18 35	

Seasons
1990

	d	h	m
Equinoxes			
Mar.	20	21	19
Sept.	23	06	55
Solstices			
June	21	15	33
Dec.	22	03	07

TABLE 5: D

Whole days elapsed from 1986 Jan. 0
to beginning of month at day 0

	1986	1987	1988	1989	1990
January	0	365	730	1096	1461
February	31	396	761	1127	1492
March	59	424	790	1155	1520
April	90	455	821	1186	1551
May	120	485	851	1216	1581
June	151	516	882	1247	1612
July	181	546	912	1277	1642
August	212	577	943	1308	1673
September	243	608	974	1339	1704
October	273	638	1004	1369	1734
November	304	669	1035	1400	1765
December	334	699	1065	1430	1795

GHA Aries and GHA and DEC of the stars in Table 6 are calculated using D in Table 5 above, as follows:

$$L = 0.9856474\ (D + d + GMT^h/24) \quad \text{where } d = \text{day of month}$$

$$GHA^\circ\ \text{Aries} = 99.3672 + L + 15GMT^h$$

$$GHA^\circ = a_0 + a_1 L + a_2 \sin L + a_3 \cos L + 15GMT^h$$

$$DEC^\circ = a_0 + a_1 L + a_2 \sin L + a_3 \cos L$$

No	Name (Mag)		GHA	DEC
			°	°
1	Alpheratz			
	α And	a_0	97.4532	29.0122
	(2.2)	a_1	0.9999637	0.0000169
		a_2	0.0059	-0.0023
		a_3	0.0007	0.0024
2	Ankaa			
	α Phe	a_0	92.9662	-42.3827
	(2.4)	a_1	0.9999663	0.0000166
		a_2	0.0071	-0.0014
		a_3	0.0004	-0.0041
3	Schedar			
	α Cas	a_0	89.4447	56.4601
	(2.5)	a_1	0.9999592	0.0000167
		a_2	0.0095	-0.0012
		a_3	-0.0004	0.0045
4	Diphda			
	β Cet	a_0	88.6446	-18.0639
	(2.2)	a_1	0.9999654	0.0000167
		a_2	0.0055	-0.0021
		a_3	-0.0002	-0.0021
5	Achernar			
	α Eri	a_0	75.0641	-57.3075
	(0.6)	a_1	0.9999758	0.0000153
		a_2	0.0094	-0.0023
		a_3	-0.0027	-0.0048
6	Hamal			
	α Ari	a_0	67.7730	23.3971
	(2.2)	a_1	0.9999602	0.0000142
		a_2	0.0053	-0.0013
		a_3	-0.0024	0.0017
7	Acamar			
	θ Eri	a_0	54.9325	-40.3594
	(3.1)	a_1	0.9999748	0.0000119
		a_2	0.0057	-0.0036
		a_3	-0.0043	-0.0033
8	Menkar			
	α Cet	a_0	53.9809	4.0364
	(2.8)	a_1	0.9999635	0.0000116
		a_2	0.0043	-0.0020
		a_3	-0.0034	-0.0001
9	Mirfak			
	α Per	a_0	48.5407	49.8135
	(1.9)	a_1	0.9999483	0.0000104
		a_2	0.0062	0.0012
		a_3	-0.0060	0.0029
10	Aldebaran			
	α Tau	a_0	30.5888	16.4839
	(1.1)	a_1	0.9999596	0.0000056
		a_2	0.0029	-0.0009
		a_3	-0.0051	0.0004

No	Name (Mag)		GHA	DEC
			°	°
11	Rigel			
	β Ori	a_0	20.9010	-8.2149
	(0.3)	a_1	0.9999668	0.0000029
		a_2	0.0020	-0.0029
		a_3	-0.0054	-0.0007
12	Capella			
	α Aur	a_0	20.4556	45.9872
	(0.2)	a_1	0.9999469	0.0000025
		a_2	0.0027	0.0020
		a_3	-0.0077	0.0010
13	Bellatrix			
	γ Ori	a_0	18.2726	6.3401
	(1.7)	a_1	0.9999625	0.0000022
		a_2	0.0017	-0.0017
		a_3	-0.0055	-0.0003
14	Elnath			
	β Tau	a_0	18.0166	28.5989
	(1.8)	a_1	0.9999551	0.0000020
		a_2	0.0019	0.0004
		a_3	-0.0062	0.0004
15	Alnilam			
	ε Ori	a_0	15.4916	-1.2077
	(1.8)	a_1	0.9999647	0.0000014
		a_2	0.0015	-0.0023
		a_3	-0.0055	-0.0005
16	Betelgeuse			
	α Ori	a_0	10.7640	7.4078
	(0-1)	a_1	0.9999621	0.0000000
		a_2	0.0011	-0.0016
		a_3	-0.0056	-0.0003
17	Canopus			
	α Car	a_0	3.4557	-52.6850
	(-0.9)	a_1	0.9999865	-0.0000022
		a_2	0.0007	-0.0055
		a_3	-0.0094	-0.0005
18	Sirius			
	α CMa	a_0	358.2345	-16.6941
	(-1.6)	a_1	0.9999698	-0.0000047
		a_2	0.0000	-0.0036
		a_3	-0.0059	-0.0004
19	Adhara			
	ε CMa	a_0	354.8482	-28.9497
	(1.6)	a_1	0.9999734	-0.0000047
		a_2	-0.0004	-0.0045
		a_3	-0.0065	-0.0001
20	Procyon			
	α CMi	a_0	344.7249	5.2644
	(0.5)	a_1	0.9999635	-0.0000083
		a_2	-0.0012	-0.0018
		a_3	-0.0055	-0.0006

No Name (Mag)		GHA	DEC	No Name (Mag)		GHA	DEC
		°	°			°	°
21 Pollux				31 Gacrux			
β Gem	a_0	343.2521	28.0636	γ Cru	a_0	271.7755	-57.0342
(1.2)	a_1	0.9999569	-0.0000080	(1.6)	a_1	0.9999601	-0.0000170
	a_2	-0.0016	0.0004		a_2	-0.0096	-0.0014
	a_3	-0.0061	-0.0012		a_3	-0.0003	0.0046
22 Avior				32 Alioth			
ε Car	a_0	333.8113	-59.4615	ε UMa	a_0	266.0092	56.0362
(1.7)	a_1	0.9999874	-0.0000103	(1.7)	a_1	0.9999707	-0.0000165
	a_2	-0.0044	-0.0052		a_2	-0.0094	-0.0015
	a_3	-0.0101	0.0020		a_3	0.0009	-0.0049
23 Suhail				33 Spica			
λ Vel	a_0	322.4982	-43.3731	α Vir	a_0	258.2543	-11.0884
(2.2)	a_1	0.9999751	-0.0000127	(1.2)	a_1	0.9999631	-0.0000158
	a_2	-0.0043	-0.0045		a_2	-0.0052	-0.0020
	a_3	-0.0062	0.0021		a_3	0.0012	0.0007
24 Miaplacidus				34 Alkaid			
β Car	a_0	321.1073	-69.6569	η UMa	a_0	252.6166	49.3828
(1.8)	a_1	0.9999944	-0.0000129	(1.9)	a_1	0.9999739	-0.0000150
	a_2	-0.0092	-0.0047		a_2	-0.0077	-0.0026
	a_3	-0.0128	0.0032		a_3	0.0028	-0.0043
25 Alphard				35 Hadar			
α Hya	a_0	317.6426	-8.5951	β Cen	a_0	248.6652	-60.3064
(2.2)	a_1	0.9999659	-0.0000136	(0.9)	a_1	0.9999484	-0.0000144
	a_2	-0.0035	-0.0028		a_2	-0.0099	0.0005
	a_3	-0.0042	0.0002		a_3	0.0042	0.0044
26 Regulus				36 Menkent			
α Leo	a_0	307.4601	12.0382	θ Cen	a_0	247.9055	-36.3020
(1.3)	a_1	0.9999629	-0.0000153	(2.3)	a_1	0.9999580	-0.0000147
	a_2	-0.0042	-0.0015		a_2	-0.0061	-0.0007
	a_3	-0.0035	-0.0013		a_3	0.0027	0.0029
27 Dubhe				37 Arcturus			
α UMa	a_0	293.6451	61.8284	α Boo	a_0	245.6107	19.2544
(2.0)	a_1	0.9999575	-0.0000167	(0.2)	a_1	0.9999687	-0.0000155
	a_2	-0.0101	0.0008		a_2	-0.0051	-0.0028
	a_3	-0.0045	-0.0048		a_3	0.0026	-0.0020
28 Denebola				38 Rigil Kentaurus			
β Leo	a_0	282.2801	14.6516	α Cen	a_0	239.7096	-60.7779
(2.2)	a_1	0.9999646	-0.0000171	(0.1)	a_1	0.9999500	-0.0000123
	a_2	-0.0053	-0.0019		a_2	-0.0096	0.0010
	a_3	-0.0012	-0.0018		a_3	0.0056	0.0042
29 Gienah				39 Zubenelgenubi			
γ Crv	a_0	275.5970	-17.4634	α^2Lib	a_0	236.8425	-15.9851
(2.8)	a_1	0.9999640	-0.0000170	(2.9)	a_1	0.9999610	-0.0000123
	a_2	-0.0055	-0.0023		a_2	-0.0046	-0.0014
	a_3	-0.0005	0.0014		a_3	0.0033	0.0010
30 Acrux				40 Kochab			
α Cru	a_0	272.9183	-63.0209	β UMi	a_0	236.6740	74.2118
(1.1)	a_1	0.9999597	-0.0000169	(2.2)	a_1	1.0000065	-0.0000122
	a_2	-0.0115	-0.0013		a_2	-0.0162	-0.0033
	a_3	-0.0006	0.0049		a_3	0.0118	-0.0046

No	Name (Mag)		GHA	DEC
			°	°
41	Alphecca			
	α CrB	a_0	225.8423	26.7599
	(2.3)	a_1	0.9999713	-0.0000098
		a_2	-0.0043	-0.0036
		a_3	0.0045	-0.0021
42	Antares			
	α Sco	a_0	212.2314	-26.4037
	(1.2)	a_1	0.9999565	-0.0000062
		a_2	-0.0032	-0.0001
		a_3	0.0054	0.0011
43	Atria			
	α TrA	a_0	207.5781	-69.0055
	(1.9)	a_1	0.9999217	-0.0000049
		a_2	-0.0070	0.0035
		a_3	0.0141	0.0023
44	Sabik			
	η Oph	a_0	201.9741	-15.7106
	(2.6)	a_1	0.9999596	-0.0000032
		a_2	-0.0021	-0.0009
		a_3	0.0055	0.0003
45	Shaula			
	λ Sco	a_0	196.2040	-37.0970
	(1.7)	a_1	0.9999515	-0.0000016
		a_2	-0.0019	0.0012
		a_3	0.0069	0.0007
46	Rasalhague			
	α Oph	a_0	195.7959	12.5671
	(2.1)	a_1	0.9999680	-0.0000016
		a_2	-0.0015	-0.0033
		a_3	0.0056	-0.0006
47	Eltanin			
	γ Dra	a_0	190.2957	51.4877
	(2.4)	a_1	0.9999858	0.0000002
		a_2	-0.0016	-0.0054
		a_3	0.0090	-0.0009
48	Kaus Australis			
	ε Sgr	a_0	183.5569	-34.3949
	(2.0)	a_1	0.9999528	0.0000021
		a_2	-0.0005	0.0011
		a_3	0.0069	-0.0001
49	Vega			
	α Lyr	a_0	180.2506	38.7674
	(0.1)	a_1	0.9999776	0.0000034
		a_2	-0.0001	-0.0050
		a_3	0.0073	-0.0002
50	Nunki			
	σ Sgr	a_0	175.7680	-26.3177
	(2.1)	a_1	0.9999562	0.0000044
		a_2	0.0003	0.0003
		a_3	0.0063	-0.0005
51	Altair			
	α Aql	a_0	161.8422	8.8280
	(0.9)	a_1	0.9999663	0.0000086
		a_2	0.0015	-0.0030
		a_3	0.0054	0.0000
52	Peacock			
	α Pav	a_0	153.2295	-56.7836
	(2.1)	a_1	0.9999437	0.0000104
		a_2	0.0042	0.0027
		a_3	0.0093	-0.0023
53	Deneb			
	α Cyg	a_0	149.1295	45.2273
	(1.3)	a_1	0.9999772	0.0000114
		a_2	0.0038	-0.0048
		a_3	0.0068	0.0019
54	Enif			
	ε Peg	a_0	133.4929	9.8081
	(2.5)	a_1	0.9999659	0.0000144
		a_2	0.0038	-0.0028
		a_3	0.0039	0.0004
55	Al Na'ir			
	α Gru	a_0	127.5264	-47.0314
	(2.2)	a_1	0.9999560	0.0000151
		a_2	0.0060	0.0009
		a_3	0.0051	-0.0036
56	Fomalhaut			
	α PsA	a_0	115.1461	-29.6984
	(1.3)	a_1	0.9999617	0.0000164
		a_2	0.0054	-0.0008
		a_3	0.0028	-0.0029
57	Markab			
	α Peg	a_0	113.3521	15.1281
	(2.6)	a_1	0.9999653	0.0000166
		a_2	0.0050	-0.0027
		a_3	0.0022	0.0010

Pole Stars

No	Name (Mag)		GHA	p
			°	°
58	Polaris			
	α UMi	a_0	65.2371	0.7983
	(2.1)	a_1	0.9992240	-0.0000137
		a_2	0.3466	-0.0022
		a_3	-0.1882	-0.0050
59	σ Octantis			
		a_0	145.3779	0.9851
	(5.5)	a_1	0.9992832	0.0000123
		a_2	0.1693	0.0043
		a_3	0.2686	-0.0033

Printed in the UK for HMSO
Dd 737368 c.30 5/85